Biocracy: Public Policy and the Life Sciences

About the Book and Author

Biocracy, a term invented by physiologist Walter Bradford Cannon, refers to the influence of biological science on society and its public policies. Beginning with the prophetic essay "Biopolitics: Science, Ethics, and Public Policy," this book addresses various aspects of the relationships among the life sciences, society, and government. Included in the topics considered are some of the more critical issues of our time: the social responses to life science innovations; health and homeostasis as social concepts; the relationship between history and biology and that between the life sciences and the law; biocratic interpretations of ethical behavior and biopolitical conflicts; and the options, risks, and international consequences of biotechnology.

Caldwell's book is a collection of articles that he wrote on this subject over a period of twenty-five years. Of the ten chapters, four have previously appeared in scholarly journals but have undergone extensive editorial revisions appropriate to this publication. The remaining six chapters have been presented at various professional meetings but have not hitherto been available in print.

Lynton Keith Caldwell is director of science policy studies at Indiana University, where he is also professor of public and environmental affairs and the Arthur F. Bentley Professor Emeritus of Political Science.

Biocracy: Public Policy and the Life Sciences

Lynton Keith Caldwell

Routledge
Taylor & Francis Group
LONDON AND NEW YORK

First published 1997 by Westview Press

Published 2018 by Routledge
52 Vanderbilt Avenue, New York, NY 10017
2 Park Square, Milton Park, Abingdon, Oxon OX14 4RN

Routledge is an imprint of the Taylor & Francis Group, an informa business

Library of Congress Cataloging-in-Publication Data
Caldwell, Lynton Keith, 1913–
Biocracy: public policy and the life sciences/Lynton Keith Caldwell.
p. cm.
Includes bibliographical references and index.
ISBN 0-8133-7363-8
1. Life sciences—Social aspects. 2. Life sciences—Political aspects. I. Title.
QH333.C35 1987
303.4′83—dc19 87-21315
CIP

ISBN 13: 978-0-367-01384-4 (hbk)

Contents

Preface

A brief indication of the author's intent may be helpful to readers of this book. My intent is to raise public awareness of a large and complex body of concerns regarding advances in biotechnology and the life sciences generally. The chapters that follow show in different ways how biology is affecting popular beliefs and behaviors and is therefore becoming a force in public policy. The term *biocracy,* first used (I believe) by Walter Bradford Cannon, has been taken to represent the power of life science knowledge in modern society. These essays were written to help the reader see the presence of this force in today's world and realize why the influence cannot be suppressed and must rather be understood. So that the reader will not be misled, I'd like to make clear what the book is and is not.

Biocracy is a collection of essays written over a period of twenty-five years, beginning in 1963 (chapter 1) and continuing to the present (chapter 10). The essays have all been updated and revised, and not one is in its original form. The earlier essays have not lost their relevance over the years because although biopolitical issues have multiplied, the related questions of values and policies remain largely unresolved. The book is intended primarily for the general reader but may also be of interest to scientist and expert. Each essay is in essence a different version of a common message: Advances in the life sciences are bringing about social changes and conflicts that, in the interest of human welfare now and in the future, should receive more serious and widespread attention than is presently evident. This proposition will be obvious to some people who see the book. As yet, however, it seems far from apparent to most people, many of whom see the issues discussed in these essays as unrelated to one another or to the encompassing larger issue identified here as biocracy.

The chapters do not form a progressive sequence of conceptual development, from first to last. Each essay stands largely upon its own argument, although some cross-referencing does occur to avoid unnecessary repetition. Certain concepts and attributions are nonetheless repeated in various chapters because they are basic to the development that is represented by biocracy. These references are relevant to different aspects of the influence of the life sciences and biotechnology. Their recurrence is therefore not repetitious, but rather a recognition of the multiple channels through which these influences reach society. In rereading these essays, which span a quarter century, I found that, in principle, no really great changes had occurred in the policy situations described. Major advances had occurred in biology and biotechnology, but as of 1987, changes in public attitudes and legal response were only beginning to appear.

Notes at the ends of the chapters have been provided where sources for verification or additional information seemed useful. For the most part the original notes and citations have been retained. They show that many of the ideas currently in controversy are not new. They do display a historical record of the "biopolitical movement" as we may identify it in retrospect. Pertinent references have been updated and expanded but left intact where little would be gained by their replacement. The separate character of the essays (or chapters) made provision of an index impractical.

This is not a book about biology or biotechnology per se. Its subject matter is their influence. The subject could be approached by several different routes (for example through adjudication; legislative action; or biopolitical issues such as genetic engineering, reproductive biotechnology, or research in molecular biology), and several books could be written on biopolicy with no great amount of overlap in illustrative material. An effort has been made, so far as possible, to avoid repeating what has already been written. References to bibliographic and periodical sources of this larger literature will be found in the notes to the chapters.

This is not a book of advocacy beyond its argument for informed attention to the social significance of developments in the life sciences. It does not fall into the error of "biologism," which is the belief that all human behavior may ultimately be traced to biology. In controversies over the uses of biotechnology, my position for the purposes of this book is largely neutral. It is not neutral with respect to the need for informed rationality and a regard to consequences in the consideration of disputed courses of action. Nor is it neutral with respect to fraudulent or demonstrably abusive use of life science knowledge. Yet there are sincere and deeply held differences among people over the ethics of biological theory and practice. This book points out the political relevance of these differences, but it is not its function to analyze them in depth nor to ascertain the right or wrong of their contentions.

I do not offer detailed description of the laws, regulations, or institutional arrangements relating to biotechnology and the life sciences. Where needed, these provisions are noted or referenced, but to treat them thoroughly would have greatly lengthened the book and distracted the reader from its main theme. Moreover, present legal and institutional provisions are likely to be short-lived, to be superseded by other regulations and arrangements in the relatively near future.

Biocracy is truly an incontrovertible fact of life. Knowledge about life will increasingly affect the actual lives of people as individuals and as societies. It is in the interest of people everywhere to recognize that life science knowledge is changing their lives and their world and to respond to this realization in an informed and intelligent manner.

Lynton Keith Caldwell

Acknowledgments

Some of the chapters in this book were initially published in journals; others were presented at meetings of professional societies but not heretofore published. Most have been substantially revised and, where necessary, updated. They are not reprints, although portions of them are partially identical with the originals. The first chapter, "Biopolitics: Science, Ethics, and Public Policy," which was written in 1963 and published in 1964, has been regarded as "prophetic" but required only minor changes to make it as applicable in 1987 as it had been in 1963. I interpret this to indicate that ideas in advance of their time in 1963 are coming into phase with intellectual developments in the late 1980s. Shortly after publication of that article in the *Yale Review* in 1964, a prominent publisher suggested that I write a book-length treatment of the subject. I did not then think that the time was right nor that I was ready. Today I believe that if the time of public acceptance of the thesis of the book has not yet arrived, we are much closer to it than we were a quarter century ago. Perhaps the book at this time may make a difference.

I would like to express appreciation to those persons who helped prepare the manuscript: Kay McCrary, Melissa Striegel, and Karen Kurfirst. Gratitude is also due the staff of the Indiana University Library system and to Robert H. Blank, Thomas C. Wiegele, and Raymond A. Zilinskas, with whom I considered many of the issues addressed in this book in a workshop on "Biotechnology: Public Policy and the Social Sciences," sponsored by the Institute for Advanced Studies at Indiana University in October and November of 1986.

L. K. C.

Biocracy: Public Policy and the Life Sciences

1
Biopolitics: Science, Ethics, and Public Policy

Some years ago a front page column of the now defunct *New York Herald Tribune* carried a whimsical description of a new science of *biopolitics.*[1] The sensitive reader might well have been as distressed as amused by the wry humor of the columnist. J. P. Miller, already secure in his reputation for social criticism through satire in *Days of Wine and Roses,* recounted an imaginary interview between an official government biopolitician and a newspaper reporter regarding the meaning of a "new science." The episode was in fact a miniature morality play—an allegory intended to pose a distasteful question in a palatable manner.

Defined by Miller, "biopolitics is the science of proving that what must be done for political reasons is biologically safe for the human race." The reported interview occurred sometime after 1971 when the collapse of the nuclear test ban treaty had been followed by a resumption of massive testing in the atmosphere and rising levels of fallout. In order to relieve popular fears, prevent panics and antigovernment demonstrations, official biopoliticians "proved scientifically that the previous human tolerances to radioactivity and all other by-products of nuclear testing, including strontium 90, had been estimated far too low." The official pronouncement "had a wonderful calming effect on the people." Public confidence was restored.

"But," asked the reporter, "suppose that an increase in bone cancer was being caused by heavy concentrations of strontium 90 in human and animal marrow?" Some unofficial scientists said so. But the official biopolitician replied that statements which frightened people were certainly not in the public interest. "Bone cancer and strontium 90 could not be linked," he declared. "The people wouldn't like it. Therefore, by

---

Revised and extended from an article under the same title appearing in the *Yale Review* 56 (October 1964): 1–16.

definition it is biopolitically impossible." Consistent in the tradition of the moralizing fable, Miller posed one of the biggest, toughest questions of our time: Are science and politics really compatible? The philosopher-dramatist with a sociological turn of mind puts the question that way. Presumably the social scientist could also—but there are difficulties. As "scientist" he finds it impractical to ask questions about the extent of man's political capacities that the present state of knowledge does not permit him to answer. Moreover, the disciplines of social science have, in the main, tacitly assumed the infinite perfectability of mankind. To hypothesize that political man cannot or will not reshape his goals and values in the light of scientific knowledge seems disloyal to the traditions of liberalism and democracy. But while the question cannot be usefully posed in absolute and theoretical terms, it is by implication being posed daily in limited and practical situations.

In the language of politics, "it is a condition that confronts us—not a theory." An explosion of biological knowledge and technology is raising questions of public policy which until recently were hypothetical, and were therefore, from a practical point of view, unreal. Whether there is, can, or should be in any sense a science of biopolitics can easily be dismissed as facetious. But the conscientious man grows uneasy when he reflects upon the mounting problems which the life sciences (in particular) are posing for political solution. There is certain to be more biology in politics, and this could mean, as J. P. Miller implied, more politics in biology.

Now twenty-five years after Miller's 1963 scenario there is more biology in politics, but the issues are different and far more numerous. In 1963 the genetic effects of nuclear fallout was the major biopolitical issue. Since then a long list of biology-related problems has been compiled, continually being lengthened by new applications of biotechnology. The nuclear threat has not disappeared, but it has become one of many biopolitical issues, some of which are of more immediate and direct concern to people, for example, genetic engineering, screening for genetic defects, and a range of new technologies relating to human reproduction. Official government is no more comfortable with the new biotechnology than it was with the fallout issue. Bureaucratic anxiety to avoid exciting the public is still present, but today even in government, there is recognition that the advancing life sciences have significantly enlarged the political agenda and the issues must be addressed in an open and informed manner.

Science—Politics and Ethics

The scientist, the politician, and the philosopher, each in his own way is confronted by the question of how political reactions to an expanding, innovating biology will affect its application to the public happiness and welfare. And unfortunately for the policymakers, happiness and welfare do not always follow from the same course of action. Yet there are urgencies in our present biopolitical state of affairs that compel a reconciliation of ethical values and scientific facts in public policies involving the biological nature of man.

To focus attention upon these policy issues the term biopolitics affords a convenient shorthand expression, and will be used here in this context and without prejudice to the separate question of whether biopolitics, specifically, is becoming a legitimized area of interdisciplinary inquiry. Biopolitics, then, suggest political efforts to reconcile biological knowledge and popular values—notably ethical values—in the formulation of public policies. It is obviously a selective focus on a portion of the larger issue of the relationship of science to society. Important developments both in science and in society justify this selectivity.

During the first half of the twentieth century, spectacular developments in the physical sciences overshadowed major but less readily demonstrable advances in biology. Moreover, the impact of applied biology upon society often occurs on a time scale that obscures its effects—at least in the early stages. The human population explosion gained momentum as public health administration and medicine began to eliminate the "natural" controls over human reproduction. The explosion of population may be as inexorable and destructive as the explosion of nuclear energy, but the consequences of the nuclear bomb are all too readily observable whereas the potential consequences of what has been called "the population bomb" have been largely inferred through the dry and less convincing medium of statistics.

Although there is widespread and profound disagreement as to its implications, the population explosion on a global scale is now generally acknowledged. There is less awareness of a concurrent explosion of biological knowledge—an accelerating geometrical expansion of knowledge, culminating long years of accumulating inquiry in the various biosciences. It is the contemporary convergence of these two explosions—people and biology—that justifies, indeed necessitates, a focus on biopolitics.

That extraordinary advances in biological science and biotechnology have occurred and are continuing is hardly debatable. To this there has been informed and responsible testimony for some time. As early as 1964 Detlev W. Bronk, President of the Rockefeller Institute, stated that ". . . we have learned more about the nature of living matter and the mechanisms of living organisms during recent years than in all prior human history."[2] And the rate of learning has accelerated.[3] The revision of man's perception of himself and of nature that the biological sciences may require could be as drastic as the changes made by the physical sciences in man's perception of the cosmos. In 1963 journalist William K. Wyant, Jr., observed prophetically that "the rough jolts of the future, in the way man thinks of himself, will come from studies done with the microscope."[4]

The more sensational speculations growing out of biological congresses make news headlines and sober editorials. Commenting on the unprecedented implications of the emerging biotechnology discussed at the 11th International Congress of Genetics, an editorial in the *New York Times* declared that "the moral, economic, and political implications of these possibilities are staggering" and then asked rhetorically, "is mankind ready for such power?"[5] In the judgment of some of the most thoughtful students of man's biopolitical behavior the answer is "No." Representative of misgivings in the scientific community is the regretful observation of Theodosius Dobzhansky that man, comprehending the meaning of his biological evolution:

> . . . should be able to replace the blind force of natural selection by conscious direction, based on his knowledge of nature and on his values. It is as certain that such direction will be needed as it is questionable whether man is ready to provide it. He is unready because his knowledge of his own nature and its evolution is insufficient; because a vast majority of people are unaware of the necessity of facing the problem; and because there is so wide a gap between the way people actually live and the values and ideals to which they pay lip service.[6]

Biopolitical Issues

Public unreadiness to use wisely an expanding biotechnology is not merely a speculative conclusion. Popular behavior and political action (or inaction) indicates prevailing attitudes toward biological realities. A cursory look at some of the current biopolitical issues suggest a mixed and contradictory picture. In each case a confrontation of biological facts, political exigencies, and ethical values occurs in the course of policymaking.

Biopolitical issues tend to fall into two general groups differing chiefly in the directness and generality of their effects. The first group may be termed environmental. Issues in this category arise when environments are impaired as a consequence of deliberate or inadvertent human action. During the 1960s the most widely publicized biopolitical issue concerned the health effects of radioactive fallout—an issue abruptly revived by the accident at Chernobyl, USSR in the summer of 1986. Meanwhile, the effects of toxic chemicals on human health and the environment emerged as a more widespread and complex set of concerns. The fluoridation controversy was an early case in point.[7] Another was the danger of chemical poisoning through pesticides, dramatized by Rachel Carson's *The Silent Spring,* which engendered controversies described as ". . . disgraceful both from the scientific and social points of view."[8]

For failure to anticipate environmental problems of scientific advances, the scientific community bears some responsibility. In a critique on environmental biology Rene Dubos took his fellow scientists to task for gross neglect of " . . . the problems posed by the response of the total organism to the total environment."[9] He argued that because medical science has not been provided with adequate scientific knowledge of "the effects of the total environment on the human condition," the potentialities of medicine for human welfare will be severely restricted until this deficiency is remedied. When scientists themselves offer no adequate explanation of the responses of body and mind to the impact of modern technology, has the politician any choice other than to "trim" biological facts to fit political circumstance? If science cannot speak authoritatively regarding the threats to physical and mental health posed by " . . . constant and unavoidable exposure to the stimuli of urban and industrial civilization; by the varied aspects of environmental pollution; by the emotional trauma and often the solitude of life in congested cities; by the monotony, the boredom, indeed, the compulsory leisure of automated work . . . ," how can the politics of these issues be guided by science? It may indeed be argued that science, and biology in particular, are providing society with a powerful array of tools and problems, but with no adequate conceptual basis for relating tools to problems in practice.

A second group of biopolitical issues are more directly and specifically physiological than environmental. More personal in immediate impact, although scarcely less general in ultimate ramification, are biopolitical issues relating to individual human behavior in the use of cigarettes, tranquilizers, narcotics, and alcohol—and extending to the biochemical control of personality. Even more personal and at the same time of greater social implications are questions of human reproduction and future populations. In addition, ethics and biology become mutually involved in the political issue of public responsibility for public health and are perhaps most starkly posed in the issue of biological warfare. In few of these areas have people demonstrated a readiness to be guided by verifiable knowledge in a search for policies equal to the problems. On many matters, inadequate as our knowledge may be, our failure to make full use of what we do know is all the more regrettable.

Biopolitical problems—particularly the major ones—grow increasingly national, international, and even global in character. The continuing flow of air and water and living organisms around the world has always tended to spread biological phenomena into any receptive environment. Modern technology multiplies and accelerates these possibilities, but it also enables us to discover and to understand the processes of dispersion and interaction. Where cause and effect relationships in these processes have become clear they have sometimes influenced political behavior as, for example, when the sciences of epidemiology and plant pathology led to the establishment of quarantines at national frontiers and were among the factors leading to international cooperation in public health and agriculture. Continuing difficulty in controlling international traffic in narcotics and the tragic consequences of carelessness in international commerce in hazardous materials underscored the lesson that wherever men go there can be no biopolitical frontiers.

Old-fashioned political nationalism has become one of the principal obstacles to biological sanity. How much positive harm or deprivation may a national government lawfully inflict upon its own people and the rest of the world in pursuance of its alleged "sovereign rights"? Atmospheric testing of thermonuclear devices has posed the question dramatically, but a list of other major biopolitical issues, current and impending, could be extended to great length and in great variety. Obvious illustrations are found in national policies pertaining to the destruction of wildlife and its habitats, allocation of water from international rivers, contamination of air and water, control of plant and animal diseases, and increase in human populations and their mass migration.

The inadequacy of conventional political mechanisms to deal with the problems of the new age of biology is nowhere more apparent than in the oceans from which life may well have come and from which man is increasingly drawing sustenance. As knowledge of the influence of the oceans upon terrestrial life continues to grow, so too does apprehension concerning impairment of their life-sustaining qualities. Massive discharge of untreated biological and industrial wastes into rivers, lakes, and coastal waters has impaired or destroyed important resources of food supply and recreation; residues from oil-burning seacraft have been so harmful to marine life that international control efforts have been sought; and proposals to bury radioactive wastes in the sea have aroused fears and controversy. But deliberate pollution is not the only problem. The Surgeon General of the United States Public Health Service reported as early as 1963 that the insect-killer DDT had invaded the water environment of the world and was being found in surprisingly large concentrations in the fats and oils of deep sea fish.[10] As of 1987 the pollution of marine and inland waters has become more complex and more serious.

But the most portentous biopolitical issues relate to the evolution of man himself. The coincident and related explosions of human population and of biological knowledge may conceivably represent the most critical state in human evolution since the last great ice age. The ability and necessity to control the numbers and hence (in some respects) the genetic characteristics of future populations could create a situation without precedent in human existence. And, in addition, the availability and refinement of chemopsychiatric drugs suggest both hoped-for and frightening possibilities for the manipulation and control of human behavior. Never before has the necessity and the possibility of control over man occurred at so decisive a conjunction.[11]

Popular (and political) "wisdom" tends to avoid facing issues in advance of a compelling necessity. Questions as sensitive and confused as those just selected for illustration are especially good candidates for relegation to some indefinite future. But if society's ability to deal effectively with a problem requires policy decision *before* the matter becomes a compelling issue then some means must be found to enable political action to anticipate the future. Practical biopolitics calls for a degree of foresight that in the lexicon of conventional wisdom would be termed *theoretical*. And practical democratic politicians find it difficult to persuade themselves or their publics of the necessity of dealing with tomorrow's uncertain problems when the self-evident issues of today press for attention.

Basis of Biopolicy

At the root of these questions one finds familiar dichotomies: fact and value, science and tradition, knowledge and action. As society moves ever more rapidly into an age of biology how well can public leadership—scientific, educational, and political—bridge the gulf between the realities of popular concepts and the realities of scientific evidence? If a massive reorientation of popular attitudes would be necessary for society to benefit fully from the *present* state of biology, how much more orientation may be required to develop a popular receptivity to the biology and biotechnology of the emerging future? There are wide gaps to be bridged between the biological sciences and public policies, and present commitments are not adequate to the task.

The building of a better bridge between science and society leads to consideration of four basic elements in the process. These are: first, prevailing perceptions of man's relation to nature; second, the meaning of science as interpreted by formalized education; third, communication between scientists and policymakers; and fourth, leadership toward a policy synthesis of scientific knowledge and ethical values. Whatever utility the concept of biopolitics possesses is primarily in relation to this fourth element. But all four are ultimately interrelated. Concern with these aspects of the science-society relationship has, during recent decades, received attention from a relatively small number of scholars, policymakers, and a few journalists. This concern has, however, had a relatively restricted impact upon public attitudes or politics generally. Political action on biopolitical issues seems to be more often forced by practical events than by foresight or considered judgment on the social consequences of life science advances.

It is commonplace that man's self-perception in relation to his environment is influenced by his cultural patterns. In cosmopolitan and dynamic societies, these perceptions may range widely, as they have for example in the history of the American people. But in the realm of politics and social policy, some perceptions prevail over others. And, with acknowledgment of the inevitable exceptions, it appears that human behavior in relation to the environment has in the main been possessive, exploitive, and shortsighted. From science, society has more often sought technology than understanding. As an eminent ecologist has said, ". . . the power of applied science has been overwhelmingly employed to exploit space, while those aspects of science that could illuminate its wise and lasting use are still largely ignored."[12]

Industrial man (which until recently meant Western man) has for the most part seen himself as separate from and outside of nature. From this inference he has frequently concluded that he may exploit nature with impunity and that where nature fails to meet his wants, science through technology will synthesize a substitute. As *The Wall Street Journal* optimistically editorialized with respect to man's insatiable needs: "Technology, as always, can serve them."[13] There has also been in Western civilization a perception of man *in* nature and a belief that he should seek understanding of his true needs and welfare through science. But this has been a minor current in a mainstream that uses science as servant rather than as teacher.

How science is used depends in large measure upon how its meaning is interpreted in the processes of formalized education. Science has been a potent influence upon education, but educational theory and practice have also shaped the course of science. The development of science depends significantly upon the nature and extent of the incentives and support accorded it in the educational structure.[14] For example, progress in fields as apparently diverse as medicine, human relations, and city planning has been retarded because of past neglect of the environmental sciences. Interpretation of the implications and the needs of science to educators and to the public at large therefore becomes a crucial element in the advancement of science as well as of society.

The expansion, specialization, and diversification of biology and of all other sciences multiplies the difficulties of communication. New sciences create a need for new syntheses to enlarge understanding of the relevance of their findings. New interdisciplinary areas take shape to deal with the new questions emerging between diverging sciences. In time many of these interdisciplinary areas develop into coherent disciplines—into new sciences—and the process of specialization and of divergent and emergent disciplines continues.

Throughout this process direct and meaningful communication between the highly specialized research scientist and the public policy maker becomes ever harder to achieve. Popularizers of science have appeared in response to public need and interest. But their status is often as uncertain as their role is difficult. The best of them find careers in journalism and may win recognition among scientists for informed and competent reporting. Scientists who popularize, however, often receive less charitable treatment from their peers. But there is presently little room for popularizers in the structure of formalized education even though the need for better communication between science and the rest of society is now widely recognized.

The problem of how to organize this communication is yet to be solved. This is perhaps because communication is not merely exchange of information. And information is itself more than mere data; it is data plus meaning—interpreted and understood. The mere possession of scientific knowledge holds no promise of its use in discovering the true needs of men or in serving the public happiness or welfare. There is need for more knowledge—but even greater need for more understanding.

Development of valid and coherent concepts of man in nature requires an interrelating and a synthesizing of knowledge. It is a task of interpretative leadership. Committees of specialists may assist the clarification and integration of knowledge, but synthesizing insights and perceptions more often originate in the minds of individuals who only then can become the expositors, the interpreters, and the advocates of a new view of man and nature. This mediating role between science, ethics, and public policy may be filled in various ways by persons from varied backgrounds—from the sciences, from professional education, from philosophy, religion, or public affairs.

Among the more effective intermediaries between science and ethics in political life have been those public servants who have in their own ways been biopoliticians in the best sense. These men and women have not only seen a relationship between scientific knowledge and the public welfare, but they have acted on this insight to influence the course of public policy. In the annals of American biopolitics one may cite Harvey W. Wiley's crusade for pure food and drugs; Hugh H. Bennett's lessons in soil conservation; Ira Gabrielson's labors to substitute science for folklore in the management of wildlife; and Rachel Carson's alerting of the public to unwanted consequences of misused pesticides. In each of these illustrations and in more that could be cited, scientific knowledge, a fundamentally ethical perception, and skill in communication were fused in effective policy leadership.

Granting that some aspects of biopolitics rest upon solid scientific support, the fact remains that we have not yet laid down a comprehensive biological foundation upon which a "science of humanity" can safely be erected. The scientific basis of biopolicy is fragmentary and will most likely remain so until the need for a comprehensive, verifiable, conceptual foundation for a healthful, creative, self-renewing society is more widely felt than it is today.

Bridging the Gaps

Better popular understanding of the biological factors in society should follow from a more accurate popular comprehension of science in the broadest sense. George Gaylord Simpson has pointed out the integrating role of biology among the sciences: "life is the phenomena to which all principles of science apply."[15] In certain specialized areas of biology, notably in relation to agriculture and medicine, there has been a continuous flow of knowledge from the laboratory to practical application. The histories of the agricultural extension service and of the public health movement in the United States afford cases in point. But the dual explosions of population and biology create a much broader need for the desirable kind of biopolitics that has been so effective in particular cases. To achieve this objective will require changes in law and may require new machinery in government. It will entail changes in the structure and content of formal education and the addition of new elements to the career development of teachers and public officials who in the long run are among the principal architects of public policy.

It is neither possible nor necessary to examine here the ways in which the machinery of government might more effectively promote and utilize scientific knowledge. Relations between science and government have been analyzed at length[16] and have been under recurring study by several congressional committees.[17] There is agreement in principle that government must be adapted to the new conditions wrought by science, but less agreement on what changes should be made. The changes in education that are needed to bridge the gap between biology and politics are more clearly evident. Throughout the modern world communication and understanding suffer greatly from gaps in the structure of education—gaps that appear with specialization and with divergence among the sciences and between them and the humanities. But even C. P. Snow's pessimistic analysis of "the two cultures"—the sciences and the humanities—does not postulate a gap that is unbridgeable.[18] The structuring of knowledge in Western society is a major factor in the cleavage that he described, but the structure of knowledge also changes, ultimately, to accommodate its content.

There is perhaps no *one* best way to obtain a more adequate communication among the disciplines and a more effective integration of related knowledge. Among the older disciplines changes in concept and emphasis may be needed as, for example, in geography where the discredited "environmentalism" of the past generation has been replaced by a search for a more valid basis for understanding environmental relationships. We will also need new disciplines to interpenetrate the older ones—to give us syntheses—to provide the form and substance of a more comprehensive understanding of man and nature. The beginnings of answers to these needs may be in new disciplines such as biometereology, regional economics, and outer space environmental research, and may be seen in numerous hybrid disciplines in the life sciences, less evident in the more slowly developing social sciences. Recently formed organizations such as the Association for Politics and the Life Sciences and the Association for Interdisciplinary Studies illustrate movement toward a reconfiguration of knowledge.

If the conditions for a better biopolitics require more realistic popular perceptions of man in nature, one way to assist this popular understanding is through the reeducation and training of teachers, public officials, and opinion leaders. Improvement of the teaching of biology has been for some time a continuing subject for attention in the various life sciences. The development of an awareness and comprehension of the significance of scientific developments by persons outside the fields of science is a different although related problem. Both developments are needed in strengthening the foundation for an enlightened biopolitics. The social and ethical aspects of the "biological revolution" have received the lesser development in colleges and universities. Efforts to strengthen education in biopolicy are now, however, underway.[19]

An important but relatively neglected avenue toward broader public understanding of science is adult education in its various forms. There is special need for an interpretation of science in its most fundamental sense to be built into career development programs for executive officials in government, business, labor, and the professions, and even more so in legal education and the administration of the judiciary. Science (and particularly the biological sciences) has heretofore received comparatively little attention in these efforts—possibly because the relevance of the sciences to most fields of career development has not been fully appreciated, again an illustration of the influence of attitudes on the place of science in society. As mankind is progressively confronted by unprecedented possibilities growing out of biological research and by exaggerated difficulties growing out of increasing populations, the need for biopolitical reorientation may gain a general recognition that it does not now enjoy. But this task of reorientation will not be done well unless the implications of biology can be reduced to terms and concepts meaningful for public policy.

To bring about an updating of the biopolitical understandings of teachers and leaders in public affairs, a valid conceptualizing, interpretative educational leadership will be needed. Some of the leadership may come, as it has, from government itself. More will need to come from the universities, the learned professions, and research institutes. For it should not be inferred that biology offers ready answers to all the problems it defines or its applied technology creates. Closing the gaps of knowledge and restructuring that knowledge for attack upon new or persistent biopolitical problems will be, even more obviously than it has already been, a continuing multidisciplinary task.

Some eminent scientists have shown skill in relating science to social needs and ethical values.[20] But these exceptional individuals have been too few and too infrequent to accomplish unaided the task of bridging the gaps between bioscience and biopolitics, between science and society. The sheer mass and specialized complexity of expanding knowledge creates a need for a systematic and continuing effort toward synthesis from which intelligible conceptualization and communication may be forthcoming. As yet the task is barely attempted and then in only a relatively few places.

A "New Science"?

Does all this then imply the need for a *science* of biopolitics for purposes quite the opposite of those suggested by J. P. Miller? The answer is both yes and no. It is probably "no" if biopolitics is understood *only* as a new formal academic discipline to deal comprehensively with social applications of biological knowledge. This is not to say that such a discipline is unnecessary, would be impractical, or that it could not be developed. We have been concerned here with the problems and the needs suggested by the term biopolitics, with general approaches to solutions rather than with specific remedial methods. But if a science termed biopolitics is not specifically implied, the need should be evident for a more effective relating of the biological to the social sciences and of both to public policy and ethics.

Without the interrelation and distillation of scientific findings into issues amenable to political action, the gap between science and politics cannot be successfully bridged. Science as technology may be readily available to the lower and more routinized levels of administration. But at the higher executive and legislative levels of the governmental hierarchy where the broad public policies are formulated, the science most relevant to the issues will be more conceptual than technical. The impact of scientific thought upon public policy will in large measure depend upon its being expressed in terms meaningful to political and administrative practitioners. The legislator, judge, and public administrator must make their own policy syntheses, but they can do their jobs more effectively if the data relevant to these decisions has been organized and reduced to understandable terms. The case for aid to the administrator in his task of synthesis has been stated with exceptional clarity by a public official and educator, Paul H. Appleby:

> Specialist after specialist pursues analysis; who pursues synthesis, or even pursues analysis with any sensible orientation to the larger function of synthesis? It is the synthesis which involves all the heavy burdens of practitioners, and these burdens are heaviest when the social-action is most complex and most complexly environed. Synthesis becomes more and more important as one goes up the hierarchy, and more and more important as one moves from the relatively specialized fields of private administration to public administration.[21]

This synthesis of itself does not necessarily require new sciences. Ecology, for example, has long been an established if insufficiently utilized "organizing" science. Its further development and involvement with the social sciences could provide much of the needed synthesis. It also seems probable that new emphases will emerge in established disciplines; that interdisciplinary studies will increase; and that new formalized disciplines may emerge.[22] New arrangements to facilitate interdisciplinary studies involving synthesis of architecture, engineering, public health, and natural resources administration—are already under consideration in a number of universities. The Association for Integrative Studies was founded in 1979 to promote, among other things, research and teaching through the development of interdisciplinary theory, methodology, and research design. Of more direct relevance, the Association for Politics and the Life Sciences was established in 1980 with similar integrative objectives specifically oriented toward public policy. In 1987 four members of this association at the invitation of the Institute for Advanced Study at Indiana University prepared a report on *Biotechnology, Public Policy, and the Social Sciences: Critical Needs in Teaching and*

Research.[23] From these developments may come significant contributions to the conceptualizing and formulation of academic and public policy in the years ahead.

Biopolitics therefore suggests a need that may be met in many different ways. It would be difficult to argue that existing educational arrangements are adequate. But because few educational needs can be shown to be fully served, the question will be asked: How important is this underdeveloped area of biopolitics in relation to other unfulfilled educational demands? Restating biopolitics in broader terms as "the study of the role of science in society," its priority is of the highest. We have been paying heavy and steadily rising prices in dollars, health, and happiness for its relative neglect, and have entered, inadequately prepared, upon a decisive test of our capacity to avoid becoming the victims of our own ingenuity.

The "condition that confronts us" calls for more than the mere tolerance of imaginative innovation in reshaping and accelerating the education of society. Tangible and timely encouragement is needed for pathbreaking efforts, for the ever risky tasks of synthesis, for the continuing development of creative individuals capable of conceptualizing and interpreting the issues that arise at the meeting point of science, ethics, and politics. The study of biopolitics—whatever it may be called—requires an extraordinary fusion of understanding, audacity, and humility. The challenge is matched by few equals.

Notes

1. J. P. Miller, "The New Science," *New York Herald Tribune* (2 September 1963): 1.
2. Detlev W. Bronk, letter to the author, 5 February 1964.
3. E.g., Bentley Glass, "Information Crisis in Biology: The Exponential Growth of the Biological Sciences Has Created Problems in Education and Information Retrieval That Are Rapidly Approaching a Critical Stage . . ." *Bulletin of the Atomic Scientists* 18 (October 1962): 6–12. A quarter century since this assessment the overload of information has increased significantly, helped only partially by computerization. See also Paul R. Ehrlich, "The Biological Revolution," *Center Magazine*, November 1969, 28–31; and Donald Fleming, "On Living in a Biological Revolution," *Atlantic Monthly* 223 (February 1969): 64–70.
4. William K. Wyant, Jr., *St. Louis Post-Dispatch*, 20 October 1963, 5C.
5. Editorial, *New York Times* (12 September 1963): 36.
6. Theodosius Dobzhansky, "The Present Evolution of Man," *Scientific American* 203 (September 1960): 217.
7. Robert Crain, et al., *The Politics of Community Conflict: The Fluoridation Decision* (Indianapolis: Bobbs Merrill, 1969); Douglas Jackson, *The Social and Legal Aspects of Fluoridation: An Inaugural Lecture* (Leeds: Leeds University Press,

1967); and Anne Gotzsche, *The Fluoride Question: Panacea or Poison* (New York: Stein and Day, 1975).

8. Rene Dubos, "Environmental Biology," *BioScience* 14 (January 1964): 12.

9. Ibid., 11.

10. Luther L. Terry, "The Complex World of Modern Public Health," *Third Annual Bronfman Lecture before the American Public Health Association,* Third General Session, 14 November 1963, Kansas City, Missouri. Report in the *New York Times,* 15 November 1963, 21.

11. Cf. Hudson Hoagland, "Cybernetics of Population Control," *Bulletin of the Atomic Scientists* 20, February 1964, 2–6; Aldous Huxley, *The Politics of Ecology: The Question of Survival* (Santa Barbara, California: Center for the Study of Democratic Institutions, 1963); and Robert S. de Ropp, *Drugs and the Mind* (New York: Grove Press, 1960). More recent citations would only document the spread and intensification of threatening conditions.

12. Paul B. Sears, "The Inexorable Problem of Space," *Science* 127 (3 January 1958): 10. The history of the National Space and Aeronautics Administration in the 1980s suggest that Sear's statement now could be applicable to both outer and terrestrial space.

13. Editorial, *Wall Street Journal,* 17 October 1963, 10.

14. Norman W. Storer, "The Coming Changes in American Science," *Science* 142 (25 October 1963): 464–67.

15. George Gaylord Simpson, "Biology and the Nature of Science," *Science* 139 (11 January 1963): 81–88.

16. E.g., F. Stefan Dupre and Sanford A. Lakoff, *Science and the Nation: Policy and Politics* (Englewood Cliffs, New Jersey: Prentice-Hall, 1962); James L. McCamy, *Science and Public Administration* (University, Alabama: University of Alabama Press, 1960); Don K. Price, *Government and Science: Their Dynamic Relation to American Democracy* (New York University Press, 1954); and Anne L. Hiskes and Richard P. Hiskes, *Science, Technology, and Policy Decisions* (Boulder, Colorado: Westview Press, 1986).

17. E.g., *Science* 142 (8 November 1963): 645–50.

18. C. P. Snow, "The Two Cultures: A Second Look," *Times Literary Supplement,* London, 25 October 1963, 839–44.

19. See whole issue of *Politics and the Life Sciences* 5 (August 1986); and Robert H. Blank, Lynton K. Caldwell, Thomas C. Wiegele, and Raymond A. Zilinskas, *Biotechnology: Public Policy and the Social Sciences, Critical Needs in Teaching and Research,* A report to the Institute for Advanced Study (Bloomington: Indiana University, Institute for Advanced Study, 1987).

20. E.g., Theodosius Dobzhansky, *The Biological Basis of Human Freedom* (New York: Columbia University Press, 1956); Pierre Lecomte du Noüy, *Human Destiny* (New York: Longman's Green and Co., 1947); W. H. Thorpe, *Biology and the Nature of Man* (Oxford: Oxford University Press, 1962); C. H. Waddington, *The Ethical Animal* (London: Allen and Unwin, 1960); Van Rensselaer Potter, *Bioethics: Bridge to the Future* (Englewood Cliffs, New Jersey: Prentice-Hall, 1971); Vernon Reynolds and Ralph Tonner, *The Biology of Religion* (New York: Longman, 1983); and Daniel J. Callahan, *The Tyranny of Survival* (New York: Macmillan, 1973). For other references see chapter 6.

21. Paul H. Appleby, letter to the author, 6 May 1953; cf. Paul B. Sears, "Human Ecology: A Problem in Synthesis," *Science* 120 (10 December 1954): 959–63.

22. E.g., for political science, Harold Lasswell, *The Future of Political Science* (New York: Atherton Press, 1963) 217 ff. and a more detailed treatment, "Political Science of Science," *The Scientific Monthly* 83 (January 1957): 34–44. As in many areas of thought, Lasswell was prophetic, but many years ahead of events. For a comprehensive treatment of synthesis as an intellectual problem, see Peter A. Corning, *The Synergism Hypothesis: A Theory of Progressive Evolution* (New York: McGraw-Hill, 1983).

23. Blank, et al., *Biotechnology, Public Policy, and the Social Sciences.* See also by Thomas C. Wiegele, *Biopolitics: Search for a More Human Political Science* (Boulder, Colorado: Westview, 1979).

2
Social Response to Advances in the Life Sciences

On 28 February 1912 William Bateson delivered at the University of Oxford a prescient address entitled "Biological Fact and the Structure of Society." He began with the observation that " . . . the civilized world is at length awakening to the fact that the knowledge needed for the right direction of social progress must be gained by biological observation and experiment."[1] By profession a horticulturist, he counterposed what he believed to be the findings of evolutionary biology and genetics against prevailing social and political tendencies. Bateson was not a social Darwinist in the extreme or pejorative sense, and disagreed with Herbert Spencer's laissez-faire political economy. But he believed that the quality of human life in the long run required that social policies respect biological facts. He was an advocate of *biocracy* before the invention of the term. Although his biases were evident, he was more inclined to raise problems than to prescribe answers. He concluded that:

> Whatever is doubtful, this much I think certain, that we are fast nearing one of those great secular changes through which history occasionally passes. The present social order is too unstable to last much longer, and he must be callous who greatly desires that it should. What will emerge from the approaching histolysis no one can predict. Let us hope, something better: and to this end may those upon whom devolves the duty of rearing that new organism, which is to grow from the dissolved tissues of society, be guided in their treatment, like physicians of the modern age, not by nostrum merely, but by the facts of natural physiology.[2]

Bateson saw that an explosion of human populations coincided with the release of energy latent in the coal fields, enabling man to begin "to devour the capital of the earth."[3] He believed that industrial society was " . . . passing through a phase which is quite exceptional in the history of the species."[4] As mechanized energy substituted for unskilled labor he believed a decline in the birthrate would be the most promising omen for the happiness of future generations. But he cited the economist Alfred Marshall to the effect that if population levels continued to rise ". . . it will be impossible for the habits of comfort prevailing in Western Europe to spread themselves over the whole world and maintain themselves for many hundred years."[5]

Although he applauded the growth of humane practices in society and, unlike Spencer, believed that biological principles suggested the desirability of restraints on competition, he was also concerned that the genetic quality of society not be weakened inadvertently by misconceived humanitarian policies. His approach to eugenics was very cautious, seeing major social as well as scientific problems in improving the genetic quality of human populations. He recognized the significance of environmental factors influencing human potential, but he insisted that no matter how favorable, they could not enlarge or compensate for qualities that were not present. "Of abstract rights," he declared, "biology knows little; of equal rights, nothing."[6] He believed that the improvement of the human condition required freedom to innovate and that the great diversity among humans that had resulted from evolution was an advantage to be preserved except in the case of severe genetic unfitness. Thus he favored an open society, but accepted the legitimacy of distinct social classes. He preferred liberal democracy to socialism because he felt that the latter required centralized authoritarian controls inconsistent with creative freedom. But he regarded the equalitarian ideals associated with democracy to be inconsistent with biological "facts" as he understood them.

The influence of this lecture is unrecorded. But it brought into perspective social questions that were to gain prominence in the expanding, populous, energy intensive, innovating political economies of the twentieth century. His thesis, that human welfare requires that social policy be consistent with biological fact, is a corollary of biocracy—a proposition that the principles of law and policy governing human society should be consistent with the facts of life as revealed by biology.[7] How to persuade a critical mass of people to accept this proposition would have been from his point of view a central task of policy.

Political Truth and Biological Fact

A basic tenet of biocracy is that prospects for human well-being and survival depend upon the validity of popular attitudes toward living nature, especially human nature. The term biocracy was invented to signify a sociopolitical system in which biological knowledge and technology strongly influence social behavior and thus ultimately affect policy. As used here it does not imply an arbitrary subordination of social and political values to the presumed facts of biological sciences. Its elementary postulate is that if and when a biological finding or fact has been demonstrated to be as consistent with reality as any finding can be, political or moral propositions that are inconsistent with it are, at the very least, questionable. This assumes, of course, that the facts of biology may be established more empirically and reliably than those moral, aesthetic, or preferential judgments that are often treated as political facts. But facts as understood in politics and science often represent different and disconnected realms of rationality. For example, the United States Congress might establish by law that the human person begins life at conception and hold, as it has, that marijuana is a dangerous substance and tobacco is not. By law these judgments would be political facts, but from a biological perspective they would have no scientific factual basis. Under biocratic politics they would be regarded as lacking rational significance.

The term fact as used here therefore implies no more than a proposition believed to be true. The point at issue in comparing social response to scientific fact is the rationality of the basis of belief. The question of rational propriety becomes political when, for example, the kind of evidence called "scientific" is proposed as the basis for policy decisions (e.g., discouraging smoking, encouraging fluoridation of drinking water, or mandatory testing for AIDS or illegal use of drugs) as against the belief that personal habits, privacy, or preferences are inalienable rights. A question not at issue here is whether scientific methods and findings are more rational (i.e., reasonable) than other ways of understanding. One may believe that the methods of science offer more reliable approaches to understanding reality than, for example, revelation, and yet accept the possibility that there are ways of knowing different from science and not accessible through conventional scientific inquiry.

The strength and persistence of popular attitudes and beliefs does not necessarily depend upon their intrinsic truth. The attitudes that have shaped politics often have been contrary to the realities of nature. Nevertheless, a popular belief remains a social "truth" of which successful politicians take cognizance. The actual truth or falsity of the belief may have no political significance *unless* there is a prospect that the force of experience, or of new knowledge, will cause people to unpredictably revise their opinions. The possibility that today's popular truth may be displaced tomorrow is one of the uncertainties with which politicians live and which doubtless accounts for some of the ambiguity of political rhetoric.

The arrival of the biological and behavioral sciences to prominence in modern society has induced a chain of consequences which continues to lengthen and proliferate. The initial effect of these sciences upon public opinion has been a mixture of positive and negative response. In its early states, the effect of biobehavioral science was more often to discredit old beliefs than to provide guidelines toward health, sanity, and survival. Some guidelines emerged, notably in public health services and sanitation. Yet our knowledge of humans in relation to their environments remains rudimentary as compared to what it needs to be if modern society is to maintain itself, to improve, and to survive.

Since publication of Charles Darwin's *On the Origin of Species by Means of Natural Selection* (1859), biology has been a focus of moral indignation. Reactions of indignation have been widely distributed throughout society. When zoologist Alfred C. Kinsey turned his attention from gall wasps to sexual behavior in the human species he aroused a storm of controversy and very probably contributed to changes in social attitudes that had political consequences. Although no direct connection exists between the publication in 1948 of Kinsey's *Sexual Behavior in the Human Male* and the deliberate selection in 1978 of a homosexual male for appointment to the city council of San Francisco, the events are not unrelated.[8] The demoralization of sexual behavior, an emphasis on individual human rights, and a solicitous concern for a symbolic representation in public life of politically approved minorities caused the unthinkable of 1948 to become the expected in 1978. Moral conservatives may have been displeased by this test of political suitability in San Francisco, but they may find negative satisfaction in the near certainty that political liberals will shortly be enraged when biology or behavioral science punctures a cherished humanitarian balloon.

It is questionable whether biology and behavioral science have as yet contributed significantly to human wisdom regarding sex or any other aspect of life. But clearly they have undermined traditional beliefs and inhibitions. In effect, science has dethroned the old authorities, but has not replaced them with more credible successors. As authority disintegrates, society initially becomes more permissive and equalitarian. It may also become more fractional and even polarized; not everyone accepts the new dispensation of amoral freedom as socially desirable. Traditional morality on the defensive may be invigorated. On those matters of fact where the voice of science is unintelligible, whoever sets the tone of the news media, influences judges and the clergy, or controls the police enjoys an advantage in defining the truth. The traditional authorities in society—mystic, ecclesiastic, or secular—have with few exceptions been interpreters, or surrogates for the interpreters, of the natural order. As official custodians of knowledge, revealed and unrevealed, authorities of the day have often tried to suppress competitors. In so doing these dignitaries have acted as if they were also custodians of communal morality, and have marked as antisocial the behavior of persons propagating heretical notions.

In contemporary liberal, pluralist democracies authoritative interpretation of knowledge is diffused, with credence often given to the specialist in a white lab coat. But competition among ideas being widely regarded as desirable, science-based propositions that offend particular social values or ideologies are challenged. Advocates of competing interpretations of scientific findings attempt to sway the opinions of an unspecialized public that has little basis for judging between competing claims. Opinion may be distorted but is seldom successfully suppressed; dissenters today are not heretics.

Debate over the uses of nuclear energy affords copious illustrations of public confusion over who are the appropriate authorities for particular aspects of an issue. What scientists speak with authority on nuclear fallout? Do we believe nuclear engineers, physicists, geneticists, ecologists, physiologists, or theologians? On nuclear issues the public has sought assurance from scientific authority because of the mysterious and unprecedented phenomena of man-induced nuclear radiation. Most biological facts, however, appear less exotic; their application in human society touches upon phenomena (e.g., birth, death, sex, mental capacity, personal habits) that are familiar, that people think they "understand" and that impinge upon socioethical values and beliefs.

A critical problem in clarifying the relationships between science, morality, and politics is the criteria for credibility. What are the tests of truth? In traditional societies, community consensus, or "common sense," was a standard test. But in cosmopolitan, constantly changing mass societies, the consensual and even the logical or experiential basis for common sense diminishes. Wisdom is no longer associated with age; and what was regarded as true or right yesterday may be regarded as false or wrong tomorrow. And science, unlike traditional common sense, is seldom the sense of a community. However true its facts, they may be as unacceptable as Darwinian evolution is to "original creation" religious fundamentalists.

Even if scientists could agree upon their facts, those facts would still be selectively emphasized and interpreted. One may easily find sound and respected scientists who are orthodox Christians or philosophical Marxists, and who appear to have little trouble in adapting their interpretations of scientific data to fit their ideological predispositions. And such scientists, perhaps unwittingly, may become enemies of freedom of inquiry and scientific investigation. A scientist of high competence, prestige, and persuasiveness can more effectively distort or suppress the implications of scientific findings than can scientifically unsophisticated religious fundamentalists. Yet facts, however honestly and forthrightly conceived and presented, have limited power to establish themselves as truths in the minds of men. No one has more eloquently set forth the limitations of facts in the establishment of belief than John Henry Cardinal Newman who wrote of science:

> We have to take its facts, and to give them a meaning, and to draw our own conclusions from them. First comes knowledge, then a view, then reasoning, and then belief. This is why Science has so little of a religious tendency; deductions have no power of persuasion. The heart is commonly reached, not through the reason, but through the imagination, by means of direct impressions, by the testimony of facts and events, by history, by description. Persons influence us, voices melt us, looks subdue us, deeds inflame us. Many a man will live and die upon a dogma; no man will be a martyr for a conclusion.[9]

The Relevance/Irrelevance of Nature

In his book on *A Framework for Political Analysis* (1965) David Easton observed that "the idea of natural systems can in reality serve little theoretically useful purpose."[10] He reasoned that it may not be easy to determine in what sense a set of relationships constitutes a "system" and in what sense an identified system is "natural." He concluded that the search for so-called natural systems still leaves us " . . . with the crucial problem of establishing the interconnectedness of a set of variables and demonstrating their theoretical irrelevance or meaningfulness."[11] Nevertheless, Easton believed that the systems approach suggested a useful way of analyzing social behavior. In brief, contemplation of systems in a generalizing or abstract sense does not contribute greatly to understanding social behavior, but analysis of the systemic interrelatedness of specific social groups may tell us a great deal about the factors that shape their behaviors.

It is necessary, however, to distinguish between systemic relationships which man discovers in nature, and personal systems of thought or concept invented by man as aids to explanation. It was against categorical systems thinking that French physiologist Claude Bernard warned when he wrote that:

> One of the greatest obstacles to the free and universal movement of human knowledge is the tendency that leads different kinds of knowledge to

separate into systems. This is not a consequence of things in themselves, because everything in nature is connected with everything else and nothing should be viewed in the isolation of a system; but the feeble yet dominating tendency of our minds leads us to absorb other kinds of knowledge into our personal systems.[12]

Determining the relevance of nature to human affairs is partly a matter of definition. It is also partly a matter of understanding systemic functions and relationships, for as Easton observed, the mere fact that various aspects of life are systematically organized is not particularly enlightening. Any individual or entity is part of many different systems. But neither persons nor groups are necessarily aware of all of their systemic relationships and dependencies. Systems failure, as when electric lights go out or water faucets run dry, are unfailing stimulants of systems awareness. Thus, awareness of the systemic organization of the world provides a useful framework for understanding the relationship of man to the rest of nature, but its usefulness is limited by the state of knowledge regarding actual relationships among particular systems and their components.

A major deficiency common to most nature-inspired models of society has been insufficient knowledge of behavioral causes and of interrelationships among functions. Natural principles have been deduced from organically structural similarities, and behavioral analogies have been inferred without sufficient attention to the possibility of significant functional differences. Organismic theories of society are familiar historical examples. Bernard's warning is that it may be erroneous to categorize entities and to invent systems without adequate understanding of the natural relationships that an artificial system may attempt to simulate.

Natural *processes* have often proved easier to model than whole systems; their sequences easier to specify, their detailed functions more adequately measured. However, the dynamics of physical systems—homeostasis and cybernetics—have been described, sometimes in relatively precise terms. The usefulness of these models to understanding of social systems is limited. Their value is indirect, but heuristic—which means that they may help us to discover what we might otherwise miss. Human life depends upon continuing interrelationships with physical and biological processes—notably dynamic ecological relationships—that must be maintained if life and society are to endure. Beginnings have been made in the modeling of social behavior, but there is as yet much disagreement concerning the validity and reliability of the models over their usefulness as guides to policy.[13] The ability of modelers to simulate the scope, diversity, and dynamics of socioecological systems has been questioned, notably in those efforts to simulate man-environment interactions on a global scale.[14]

Debate over the relevance of nature to humanity has often been too narrowly defined to draw out the truly important issues. Dogmatic opinions regarding man-nature relationships have led to claims and counterclaims that miss the point that man is part of nature and does nothing that cannot be done within the range of possibilities in the natural world.

While much of the argument that nature is irrelevant is refutation of premature or ill-founded claims for relevance that may or may not exist, there is more basic resistance to "naturalism" in both popular attitude and philosophic thought. This resistance stems from the belief that the important things in life which man presumes to control have little to do with nature. Thus, qualities which men have sometimes attributed to nature, such as frugality, bounty, beauty, harmony, and ferocity, are inherent in acquired human perception—not in nature. The truths of natural science are not exceptions. Not only are scientific facts perceptions conditioned by the limited sensory equipment of human physiology, but man has also claimed knowledge of higher truth above and beyond nature and which has often been preferred to the truths of science.

There is no need to labor the case that "nature" as recently understood in modern Western society has been widely regarded as a manifestation of God's plan for mankind, and from this perception rationalization for both exploitation and stewardship have been derived.[15] Nevertheless, there have been serious efforts to discover a social reality common both to natural science, religion, and ethics. The authors of these efforts do not regard nature as irrelevant, but do not necessarily agree on the extent of this relevance and on what it implies for human welfare and survival.[16]

Social values embodied in religion are not necessarily incompatible with scientific theory and evidence, but some religious beliefs and dogmas clearly are. We can hardly dispute Cardinal Newman's statement that whereas facts come from knowledge, principles come from faith; nor prudently ignore the implications of his question: "Why are we so constituted that Faith, not Knowledge or Argument, is our principle of action?"[17] Behavioral science has yet to provide a generally accepted answer. One may of course have faith in scientific knowledge and act on that faith. This occurs daily for millions of people in scientifically advanced countries. Nevertheless, some comparatively well-informed persons have preferred scriptural to scientific explanation of the facts of life. And throughout Western technoscientific society there continues to flourish massive popular curiosity regarding wholly unfounded interpretations of natural events.[18] It is difficult to know whether the popularity of notions such as the chariot of the gods, the Bermuda triangle, pyramidology, scientology, and worlds in collision represent genuine belief, or whether they merely intrigue the fancy of people poorly schooled in science, unsophisticated in religion, and responsive to entertaining speculation.

The apparent irrelevance of scientific facts and theories to the routine concerns of many people lies in the way in which facts reach or fail to reach public awareness. Communication requires receivers as well as senders and it is commonplace that the messages sent are not necessarily the messages received. All information received is filtered in the mind of the receiver. It is no contradiction to observe that people *know* facts that they do not *believe*. For facts to alter folkways they must reach the minds of people in the forms that they can process and internalize in their behavior. Advertisers of commercial products resort to applied psychology when they try to associate their message with the broad range of relationships, belief patterns, and values closest to the customer's innermost feelings. Thus the advertiser endeavors to place his facts in a matrix of accepted belief and behavior. But this matrix of popular attitudes often contains many traditional or conventional elements that do not easily accommodate scientific facts which in this context are therefore perceived as irrelevant. For them to gain relevance, the matrix or belief pattern itself must change. But the purveyors of scientific information are usually specialists, prepared to argue for the truth of their conclusions, but often unable to relate their message to the life experience of people for whom the message is intended.

Relatively few scientists seem prepared to explore an integration of their conclusions with those from other disciplines, to join other specialists in translating analysis into synthesis, in assisting the construction of facts and concepts originating within their disciplines into a more general social theory or ideology. For many scientists this would be regarded as none of their business. For a prestigious natural scientist to attempt this would not advance his special competence, and could hazard his reputation. It could raise doubts amongst his science peers and invite rejection from social scientists who did not find his credentials convincing evidence of competence as a social theorist. Nevertheless, some prominent natural scientists have taken excursions into social policy with varied results (e.g., the Pugwash Conferences). And it has been easier, and more generally acceptable for them to do this than the reverse case of social scientists attempting to contribute to theory building in the natural sciences.

In short, the construction of coherent social and political explanations is not a task for scientists alone, although some have attempted it. Long commitment to the methods of reductionist science may tend to unfit persons for building hierarchies of concepts and propositions. The synthesis required to make interrelated scientific facts acceptable and believable to society, and capable of being acted upon by governments, would very likely be vulnerable to criticism in detail by science specialists, including many social scientists.

It is not that the sciences, natural or social, are inherently unable to bridge the gap between scientific information and its social perception. But the effort would be time-consuming, often frustrating, and always arduous. Rewards to the participants would be uncertain. For such multidisciplinary/interdisciplinary tasks, there are no readily identifiable peer groups to provide review and competent criticism. Potential sponsors are therefore wary, reluctant to provide funds and facilities for efforts not greatly in demand and the outcome of which would be uncertain. Thus the work of educational synthesis proceeds slowly against resistance.

For nature to be relevant to the lives of people its propositions must be integrated conceptually into the lives that people actually experience. People must be able to see how biological science relates to and explains the world they know. It is important to the social response that this relevance be made clear for all of that personal world to which it applies and not merely to part of it. And yet personal horizons may often be too confining to accomodate biological evidence of importance to people who in their own interest ought to understand it. Thus enlargement of personal horizons must often accompany appreciation of biological facts. Unfortunately, reductionism in science is paralleled by a different kind of reductionism in the popular perception and formulation of social issues. Common to both is a concentrated convergent focus of concern. Although these aspects of narrowed focus have no causal connection, reductionist science, as such, does not contribute to synthesis in social policy to which its findings may be related. Thus, certain theories and practices associated with the biological sciences have become the focus of single-issue politics.

On issues such as abortion and contraception, social response has tended to be unidimensional. The full context of issues relating to procreation has not as yet become a part of socioecological reality in the greater part of modern society. Sexual reproduction is widely regarded as a personal and individual affair. Illustrative of the narrowed perception of dogmatic antiabortionists is their insistence on the moral necessity to respect the sanctity of life (human) by noninterference with natural processes (of procreation). But biomedical science now makes possible the practical application of a counterethical conclusion, which is that, if "a human life" is the most precious of all things, the greatest care should be taken in bringing it into the world. Antiabortionist reasoning could be turned into argument for licensing human reproduction.

A significant effort to narrow the gap between social convention and the life sciences is not likely to occur unless it is seen as answering to a need approaching urgency in public policy. In any general sense the need has not yet been so perceived. Inquiry in so broad and complex a field of investigation must necessarily be interdisciplinary and cannot be expected to yield rapid results. The road toward more adequate understanding of the social implications of biology is an ideological and semantic obstacle course. The presence of these obstacles appears to be one of the reasons for reluctance among research funders and academic researchers to embark upon a road in which the hazards are more readily apparent than the benefits. Yet until these obstacles are identified and somehow surmounted, the gap between biological knowledge and social understanding is likely to widen, with results helpful neither to society nor to the pursuit of the sciences of human behavior.

The "Injustice" of Nature

Social response to the events and processes of nature are consistently colored (1) by an anthropocentric view of nature and (2) by misunderstanding or no understanding of how the world of nature works—man in relation to it included. Thus we hear of treacherous viruses, merciless droughts, cruel winters, and savage beasts. Nature may be "beneficent," but is as often believed to be "unfair." But although poetic license is acceptable as art—it is not a reliable substitute for understanding—and is hardly a contribution to rational policy making.

Philosophic concepts are not generally regarded as poetic—nevertheless, they often express aesthetic or ethical beliefs, judgments difficult to authenticate by reference to empirical data. For example, justice is a normative ethical value, not an attribute of nature or amenable to scientific inquiry. Hence the nonjustice and amorality of nature may be regarded as fact, but it is not a fact acceptable to many believers in anthropocentric ethics. Inconveniently, there appears to be no intermediate or neutral ground between the concepts of justice and injustice in the English language. No distinction is commonly made in political rhetoric between the absence of justice and positive injury; although, in law, significant differences may exist. It is nevertheless possible to conceive events that lack the quality of justice, but are not unjust in the sense of being morally wrong. But whether the forces of nature are merely neutral in relation to human values or whether they are regarded as sometimes positively beneficial or inimical is a matter of social attitude. Historical human societies have responded in many different ways to perceived truths of nature. Traditional attitudes have tended to be those of acceptance—propitiatory acts being reserved for abnormal or catastrophic events. Primitive and pastoral man was seldom in position to do other than conform to natural cycles and adjust to the apparent vagaries of nature.

The concept of an unjust nature is associated historically with the emergence of abstract thought in transcendental religions, and of "higher" ethics. Earlier geocentric religions of oriental and classical antiquity and of pre-Columbian America seldom viewed the relationship of nature to man in terms of justice or injustice. Nature as an expression of the will of the gods could hardly be unjust. The acceptance of tragedy as inherent in life and earth gave ancient religion a character described in Western perspective as "fatalistic." The translation of tragic to evil appears to have been formalized by the Zoroastrians, infused into Judaism and thence into Christianity.[19] But Christianity was accepted by or imposed upon peoples whose cultures were not congenial to the Judaic perspective. Thus some versions of Christianity, notably the Hispanic, retained a tragic sense of life that has historically inured it to suffering and strongly biased its understanding of the tragic as distinguished from the unjust.[20]

Social response to biological science is conditioned but not wholly determined by the prevailing ethical orientation of society. In the Western society to which William Bateson's lecture was addressed, a conjunction of developments in science, industry, economy, and social ethics had created a paradox that advances in knowledge, including advances in biology, opened the way to great improvements in the human condition, but were also being applied in ways that threatened to make things worse. The opinion of Bateson, shared by a succession of observers equally out of step with conventional attitudes, was that social resistance to the findings of biology deprived society of the benefits that might ultimately be crucial to its survival.

For persons of conventional attitude, and suspicious of science except when "useful," there is seldom occasion for inner moral conflict regarding the implications of science. Morals, conventionally interpreted, commonly take precedence over science as conventionally understood. Findings of science of which people disapprove (but cannot ignore) can be deprived by law of political legitimacy. For many biological findings, an accommodation with society has been achieved and is reflected in what the law does, or does not, say about propositions based upon science. For example, regard for the facts of epidemiology permit public imposition of compulsory vaccination and quarantine. With some conspicuous exceptions relating to human reproduction, legislative control of science today is less concerned with findings—more with methods. Controversies over cloning, recombinant DNA, and experiments using human subjects, are essentially arguments over method but with apprehension regarding results. Antagonisms over the truth and meaning of biological facts are nevertheless widespread. Facts are in dispute in such familiar issues as abortion, contraception, sterilization, euthanasia, alcoholism, cigarette smoking, fluoridation, pesticides, nuclear radiation, and intelligence testing. In these and other issues there are strongly held differences of opinion over what are to be regarded as facts and what alleged facts imply.

There is disagreement, moreover, regarding *who* is entitled to define the facts. For each of the issues just listed it is doubtful if a majority of people anywhere (possibly excepting highly literate populations of northwest Europe) would give biologists the benefit of whatever uncertainty they might have. In the United States conventional authorities in relation to issues growing out of the life sciences appear to be scientists, physicians, lawyers, clergymen, the police, and, increasingly, commentators in the television media. Opinions of persons representative of each of these categories are influential in different ways with different sectors of the public, and the nature and extent of their influence probably varies with the issue. To these bellwethers of opinion should be added professional moralists and human rights leaders who may most realistically be understood as secular clergymen. It would be to the advantage of the public for persons responsible for public policies involving the life sciences and in professions such as medicine and law in which biopolitical issues are especially significant to have better understanding than is now common of the locus, basis, and significance of social authority of biopolitical issues.

The most paradoxical social responses to biological fact may be found among the biologists themselves. It was among biologists that controversy over recombinant DNA originated and among whom the acrimony over sociobiology has been sharpest. This circumstance is partly analogous to the controversy among nuclear physicists over the development of nuclear weapons and reactors. But the substance of biology, being life itself, touches more deeply than do other fields of science, not excepting the behavioral, which may be regarded as a specialized extension of biology. And science in relation to all knowledge is not holistic; it is a particular way of learning and thinking. A scientific or technoscientific culture is now superimposed upon the traditional cultures of modern nations. Unlike the pre-science cultures of traditional societies, science does not cover all aspects of life, nor those which to most people are the most important. Like all other people, a person who becomes a biologist is born into and grows up within a particular cultural environment. And it is evident from what we know of the lives of scientists that the cultural and ideological commitments of their early years subsequently influence their inclinations, even when expressed in the name of science. For example, Harvard University evolutionist Stephen

Jay Gould is reported to have said that he "learned his Marxism literally at his daddy's knee."[21]

Controversy among scientists over what is to be regarded as fact in science has extended beyond biology; but moral and ideological overtones are especially resonant where human biology and behavior are concerned. Among scientists caught in latent conflict between fact and value there is often evidence of inner tension expressed in passionate defense of personal compromise. This compromise has in many cases taken the form of distinguishing between "good" and "bad" science. It sometimes appears as if the methods and content of which a scientist approves is good science and that science that points toward a conclusion of which the scientist disapproves is bad. But the scientific objection is usually directed toward method rather than directly against substance, although the substance may be the real target. Scientists are as able as lawyers and clergymen (among others) in rationalizing prejudices and in invoking high-principled, theoretical propositions to put down conclusions that they find personally objectionable.

It may be that persons who seek a more rational social attitude toward reality expect too much of scientists and too little of society's moral leaders. When scientists are asked to declare themselves on matters of social ethics beyond their fields of expertise, they are in effect encouraged to risk role confusion. Competence in a science does not preclude competence in social and ethical thought, but is not necessarily concomitant with it. Neither does it preclude it, as some biologists have given serious and continuing attention to the social and ethical implications of their work, and of science as an intellectual enterprise.[22]

Those aspects of nature that bring pain or unhappiness to life, that diminish the quality of life as men perceive quality, might more safely be regarded as tragic rather than, a priori, as unjust. Tragedy, like injustice, is an evaluative judgment. But it is less a moral conclusion than a recognition that man, while able to envision a world that is preferable to the one nature has provided, must nevertheless under protest and with regret accept the reality of nature's world. Under the Judeo-Christian ethic, toleration of injustice is wrong, but under that ethic the tragic has been tolerated in the name of justice. Thus hopelessly handicapped persons may be born into the world because it would be unjust to deny their progenitors the "right" of reproduction. Eugenics as science is widely rejected today, not only and probably not primarily because our understanding of hereditability in humans is inadequate to support theoretical conclusions, but because of fear of its social and political misuse. But criteria for determining what use is "misuse" appears to be often derived from nonscientific social philosophy. This is no argument for rejecting moral criteria, but it may be argued that the consequences of applying the criteria should be open to examination by methods derived from science as well as from philosophy.

The argument here is not that science should provide the foundation for social policy, nor that existing science is adequate to do so. The argument is that moral philosophy should not be invoked to obscure or distort realities that science reveals—realities with which humanity must learn to cope in the interest of its welfare and survival. Justice metaphorically may breed tragedy as well as unfitness where inspired revelation is inconsistent with the way the world works.

The Imperatives of Nature

A motto of the Stoic philosophers of classical antiquity was *Secumdum Naturam Vivesi* (To Live According to Nature). Similar attitudes have been expressed in the ethical traditions of China and Southern Asia, and appear in the moral precepts of less advanced cultures almost everywhere. For primitive societies such precepts make a virtue of necessity. For more advanced societies the thought may be honored, but practice is often indifferent. As people develop artificial systems and technologies it becomes more difficult to determine when they are living according to, or in accord with, nature, or whether their behavior is ultimately inconsistent with "natural laws." And science, contrary to what one might expect, does not resolve the question because it may often be read to more than one conclusion regarding man's relationship to nature.

In contemporary cosmopolitan society, one may find a great diversity of attitudes regarding man's relationship to nature and the uses of natural science in social policy. Generalizations regarding trends of thought and clusterings of attitudes are of course abstractions, and must be understood as simplifications of reality to facilitate comprehension. The point has been made that the "facts of life" as derived from science are indifferently received or may be rejected by many people. In technologically advanced societies, however, the predominant attitude toward science—although not necessarily toward nature—is affirmative. But these affirmative attitudes do not represent a common response to scientific facts. People hold affirmative attitudes toward the sciences for differing and sometimes opposing reasons that may lead to opposing conclusions. Moreover, one may accept science in general, but reject particular scientific findings and applications.

Two general and contrasting affirmative responses toward natural science may be identified. For convenience we will designate them as *adaptive* and *directive*. The adaptive response is in the tradition of naturalist philosophy—of living in harmony with nature. The directive response has been a dominant attitude in modern industrial society and follows the aphorism of Sir Francis Bacon in *Novum Organum* that "Nature, to be commanded, must be obeyed." Science must be pursued in order to direct the forces of nature.

While sometimes intergrading, these two responses are usually associated with different aspects of culture. The adaptive response has traditionally been associated with the humanities—with poetry, painting, aesthetics, jurisprudence, and religion. The directive view contrastingly has been associated with economic affairs, with engineering, and with military science. This distinction helps to explain the dominance of the directive response over the adaptive in expansive, industrializing society. It serves the goals of economic enterprise and public administration. The directed use of scientific knowledge is the keystone of "development," a concept that may be the one normative proposition and value prevailing in nearly all countries in the modern world.

Each response has its extremists. Those of the adaptive persuasion argue for "back to nature," to cultivation of the simple life and of soft or ecologically sensitive technologies. Small group, back to nature responses to perceived ecological and biological facts are hardly novel. They belong to the more inclusive movements of protest against the economy and technology of modern industrial society. But few of these communal efforts wholly detach themselves from the technological society or from its products and values. There are often signs of self-deception among members of these groups, and their membership is characteristically unstable. And significantly, their negative response is not wholly a consequence of their interpretation of biological or ecological findings, but rather a rejection of prevailing social and economic values.[23]

More practical, although necessarily qualified examples of adaptive responses to biological facts, may inadvertently be provided by Amish or Mennonite farm communities which in relations to the natural environment have high survival potential. This example is valid only in a very limited sense as these highly traditional communities follow custom rather than science, albeit the practical effect of their style of farming and living is generally compatible with biological facts.[24]

The more generally accepted adaptive responses to biological facts do not reject industrial technology, but rather argue for its selective adaptation to human needs. This entails the ever difficult task of striking a balance, in which values and the means to their attainment are reconciled. How far and in what directions should society depart from adapting to nature to a managing of natural properties and systems? It is here, in this category of response, that the adequacy of science for guiding man-environment relationships is most clearly tested. Where science is uncritically accepted or rejected, or where an arbitrarily unbalanced view of science is held, it is difficult to test the reliability of scientific knowledge. Its adequacy remains unexamined.

Conclusions regarding adequacy, assumed without critical examination, can hardly be helpful in a rational use of scientific knowledge in an ecologically adaptive response to problems of man-environment relationships. As man's relationships to nature become more novel and complex following advances in scientific knowledge and applied technology, it becomes correspondingly more difficult to know how to adapt to the ground rules of nature—or indeed to know in the early stages when one is failing to do so. Control through trial and error cannot keep pace with science-based innovations; attempting to do so might be dangerous where the hazards of error in the uses of powers derived from science are high. And so a politics in which survival is a high priority will place a high priority upon the testing of both science and policy in relation to fundamental objectives.

One of the more elaborate and policy-oriented, adaptive responses to natural science thus far has been *Blueprint for Survival,* a manifesto published in London in January 1972, and signed by thirty-six British scientists, physicians, and environmental scholars. The premise of the *Blueprint* was that science-based technology was being pushed beyond limits that nature, including human nature, could indefinitely support. The survival of modern man was seen to be threatened by his insistence on directing nature in ways and toward ends that would ultimately collide with natural "laws." The *Blueprint* would adapt society to ecological realities as perceived by its drafters. Its premise, stated in the following language of its preface, refers to another adaptive response which we will presently consider:

1. An examination of the relevant information available has impressed upon us the extreme gravity of the global situation today. For, if current trends are allowed to persist, the breakdown of society and the irreversible disruption of the life-support systems on this planet, possibly by the end of the century, certainly within the lifetime of our children, are inevitable.
2. Governments, and ours is no exception, are either refusing to face the relevant facts, or are briefing their scientists in such a way that their seriousness is played down. Whatever the reasons, no corrective measures of any consequence are being undertaken.
3. This situation has already prompted the formation of the Club of Rome, a group of scientists and industrialists from many countries, which is currently trying to persuade governments, industrial leaders and trade unions throughout the world to face these facts and to take appropriate action while there is yet time. It must

now give rise to a national movement to act at a national level, and if need be to assume political status and contest the next general election. It is hoped such an example will be emulated in other countries, thereby giving rise to an international movement, complementing the invaluable work being done by the Club of Rome.

4. Such a movement cannot hope to succeed unless it has previously formulated a new philosophy of life, whose goals can be achieved without destroying the environment, and a precise and comprehensive programme for bringing about the sort of society in which it can be implemented.[25]

The *Blueprint* was widely criticized as alarmist, naive, and impractical. Yet if its premises were regarded as correct (and many informed people agreed with them), it might be expected that someone would propose a specific strategy to readapt human society to the conditions set by nature. The *Blueprint* did in fact propose a road to survival through a steady-state society, one that could be sustained indefinitely while giving optimum satisfaction to its members. The strategy to this end included the attainment of four conditions:

1. minimum disruption of ecological process;
2. maximum conservation of materials and energy—or an economy of stock rather than flow;
3. a population in which recruitment equals loss; and
4. a social system in which the individual can enjoy, rather than feel restricted by, the first three conditions.[26]

To the extent that it recognized that the road to survival might lead through politics, the *Blueprint* approached realism. It proposed a Movement for Survival (MS)—a coalition of organizations concerned with environmental issues and supporting the principles of the *Blueprint*. But its implementing strategy was less realistic—and as a guide to action severely deficient. Its plan for a radical decentralization of government and the economy gave no indication of how so radical a constitutional and demographic change could be brought about. It found mass democracy to be incompetent to deal with the survival issue, yet failed to show how the mass society of present-day Britain could be persuaded to reconstitute itself into many self-governing and relatively self-sustaining communities.

In sum, *Blueprint for Survival* did not demonstrate that grand strategies are impractical, or had best not be attempted; instead it exemplified their difficulties and their hazards. The premises of *Blueprint* may ultimately be verified, and societies everywhere may one day have to live by the principles that it prescribed. But admonitions of the good, the learned, and the wise are insufficient to create a biologically rational society.

The organization called the Club of Rome, to which the preface of *Blueprint* alluded, was also a response to what its members call the *problematique.* This term summarizes the problem of what to do about modern man's commitment to "growth" in a world in which everything but man's ambition is limited. The Club's concern, expressed through several reports and conferences, has been primarily directed toward the phenomenon of growth. Although the approach of the Club to the predicament of mankind has not been primarily biocratic, it has been concerned with survival; its purpose has been to inform and to stimulate serious thought regarding the *problematique.* It has not been ostensibly a political effort, even though its members have clearly hoped to influence the behavior of governments and international organizations. The Club's early efforts and especially the first of its commissioned reports, *The Limits to Growth* (1972), applied systems dynamics analysis to worldwide trends. The analysis was, broadly defined, an effort to apply scientific methods to projections of the future.[27] These efforts, and particularly the *Limits* were sharply criticized both as to method and as to findings. Among other things, the critics objected that technological ingenuity had been unduly discounted, that the alleged dangers were remote, and the implied remedies would be unacceptable to the poor nations of the world.[28] Subsequently, Club of Rome inquiries leaned toward the position of the critics in studies supportive of the New International Economic Order.[29] But in these efforts economics and ethics provided the base of argument. The bioecological factors in the human predicament were not dismissed, but they were discounted. The adaptive response to the biological aspects of an impoverished world appears to be to place social justice and economic redistribution first upon the list of behavioral changes needed. Bioecological adjustments—such as declining birth rates in poor countries—were expected to follow automatically.

Let these examples suffice for the present, as a selective illustration of adaptive response.

The more aggressive directive response to nature and biological fact presents a greater cause for concern in a politics of survival. The root of this concern is uncertainty over whether human capacity for foresight, cooperation, and self-control can match man's ability to invent and deploy technologies of ever-increasing power. Since Bacon's time, Western society has developed under the implicit premise of unlimited human creativity. Because the limits to human potential have not always been made explicit by nature, people have often assumed that no ultimate limits exist. Human success in surmounting limits once thought to be inexorable has provided a latitude for optimism regarding man's ability to direct nature that has sometimes been carried to extreme confidence and even arrogance.[30] For this posture the Greeks had a word—*hubris*—pride unwarranted by reality.

The outer limits of optimism regarding man's directive control of nature may be found in the "conquest" of outer space. Obviously, any expedition beyond the atmospheric limits of the earth requires strict obedience to the laws of nature as margins for error diminish. Advocates of space colonization and of manned space flight cannot be either ignorant or contemptuous of scientific facts, including biological facts pertaining to life in extraterrestrial environments. But their appreciation of nature may nevertheless be unbalanced, particularly as regards the long-term and behavioral consequences of life in outer space. There are unsolved problems of technology and biology in relation to living in outer space environments, but the limiting problem may be human psychosocial adaptability under conditions differing radically from those within which mankind evolved.[31]

Requirements for survival in outer space environments impose constraints upon human behavior to which men have never conformed upon earth except for relatively short intervals or under special circumstances (e.g., hospitals and prisons). Life in submarine environments or in airplanes does not simulate social life generally. In these extraordinary environments there is little or no room for life's eccentrics, misfits, or uninhibited individualists. Life in a spaceship must be simultaneously technocratic and biocratic. The necessities of technical and biological systems maintenance absolutely must rule. Classical concepts of autocracy, democracy, and inalienable rights are as irrelevant as anarchy.

Gerard O'Neill's space habitat concept illustrates an application of scientific facts to human ambition that is rationalized beyond the practical realities of social adaptability as we presently understand it. O'Neill has proposed construction in outer space of cylindrical habitats capable of accommodating as many as 50,000 permanent inhabitants within a simulated terrestrial environment.[32]

The technical problems of constructing and maintaining such an artificial ecosystem are probably solvable. Yet in so unprecedented an experiment with the parameters of life it seems probable that social scientists, physiologists, psychologists, and particularly microbiologists rather than physicists and engineers, should have the last word (or perhaps the first). Science fiction concerned with outer space travel and settlement seems to take human adaptability for granted. Yet physical adaptability hardly guarantees the success of extraterrestrial colonization.

As to the prospects for the continuity of life in an extraterrestrial artificial ecosystem there is no science presently capable of providing meaningful probabilistic estimates. The magazine *Newsweek* philosophized: "Nothing in history suggests . . . that humans won't simply reproduce their social miseries in space."[33] Humanity has evolved selectively in relation to an environment differing greatly from modern urban centers of technological society and differing radically from proposed habitations in extraterrestrial space. A necessary precondition to extraterrestrial colonization would seem to be adaptive bioengineering, remaking the human animal to meet the requirements of an environment for which evolutionary nature has provided no prototype. Biotechnology is presently at an unknown but probably remote distance from this capability. Robotics appears to be a better bet.

Possibilities and Conjectures

Consideration of human prospects in unprecedented environments of outer space throws into high relief a proposition upon which good people would rather not speculate. It is that man's adaptability, while exceptional by virtue of his mentality, is limited by his evolutionary past, and that there is somewhere a boundary beyond which he is incapable of the social behaviors necessary to the management of unprecedented and increasingly complex affairs. The extreme demands upon human adaptability in extraterrestrial space serve primarily to accentuate those behavioral principles and problems that make the prolonged continuity of human survival on earth a matter of uncertainty.

To believe that there are no limits to human potential is on a par with the most imaginative beliefs ever entertained in a prescientific age. But because the parameters of achievement have not been identified, some will argue that to speculate upon their existence is a waste of effort. In a general sense this may be so. Yet it may be prudent to ask what might happen if untested assumptions led society to break through unperceived zones of safety into socioecological catastrophe. How would one know when thresholds of danger had been reached?

Catastrophe has always been a popular subject, perhaps because it is entertaining and has often been all too plausible. Catastrophe implies an abrupt, radical change of state or condition with adverse consequences. In our time it has been conjectured on the assumption that human technical ingenuity will overreach human capacity for social adaptation and the management of complexity.[34] In effect, society will climb out on a limb that will not sustain its weight. The consequence could be social collapse, retrogression to a less complex sociotechnology, or extinction. The latter possibility is conceivable if the biosphere is rendered unfit for human existence. Thus trial and error may not be a safe or even feasible way to discover the threshold of complexity beyond which man cannot manage his affairs. Fatal error precludes further trials.

Among those who dissent from the fear that *hubris* will lead man to catastrophic destruction are persons who warn of disaster through stagnation if the Promethean ambition of mankind is curbed. In contrast to catastrophe, the theory of stasis has a larger if less exciting literature. The sociological theories of Auguste Comte and Herbert Spencer conjectured the evolution of society to a condition approximating stasis. But stasis and equilibrium theories, of which there are numbers of variants, have now gained a new and practical significance in proposals for creating steady-state or indefinitely sustainable societies. "Growth" in its varied and often confused meanings is becoming a major focus of social and scientific concern. This is not the place to deal with the complexities of the growth issue, but biological fact is critical to its understanding. A major aspect of the growth issue concerns the numbers and dynamics of human populations and here theories must somehow respond to biological facts.

If humanity somehow avoids the penalties of overstressing the limits to growth, does a dismal future nevertheless await it? In his book *Posthistoric Man* (1950) Roderick Seidenberg conjectured an end to history; a point in time at which the human mind will have achieved all that it is capable of achieving and will have created a completely balanced, self-renewing, perfectly adapted society.[35]

History to Seidenberg is a transition between prehistory in which man was guided wholly by instinct without language or technology, and posthistory when, through civilization and science, man has learned all that he is capable of learning. Beyond history, life would be essentially unchanging and routine, and the life of the human species for an indefinite period of geologic time would, in effect, remain timeless. Seidenberg's thesis is summarized in these words:

> Consciousness depends upon a state of imbalance, a condition of tension. It is awareness, in its widest and most intensified aspects, of that unstable equilibrium whose passing phases in the development of man we call history. Thus we are led to perceive that history itself, however inclusive we may conceive its ways, must be counted in reality as a high transitional era of relatively short duration in comparison with the slumbering eternity that preceded it or the ever more static ages that will follow upon it. Consciousness will gradually evaporate and disappear in this posthistoric period, very much as it condensed step by step into ever sharper focus during man's prehistoric era. In the ultimate state of crystallization to which the principle of organization leads, consciousness will have accomplished its task, leaving mankind sealed, as it were, within patterns of

> frigid and unalterable perfection. In this consummation we perceive the essential meaning of the posthistoric period of man's development.[36]

There is an irony in Seidenberg's thesis, for it is superficially "optimistic" in the sense that it postulates the ability of man's intellect to solve the organizational and behavioral problems that would otherwise outrun his capacity for technological innovation, terminating the advancement of civilization in catastrophe. Thus a threshold to achievement exists, but is not imposed by man's ability to control his own behavior; it is reached, paradoxically, by his success in achieving social perfection. The consequence of the domination of the rational by the irrational nature of man is catastrophe; the consequence of the domination of the intellect and of rationality over man's irrational nature is ultimate social perfection, fixed and unchanging to the end of time. Seidenberg concludes his deterministic prognosis with an assessment of the ultimate cost of perfect adjustment to reality. In these final lines he reviews man's changing perception of his place in nature from the childhood of experience with civilization to the ultimate metamorphosis beyond the end of history:

> In the course of his development he has been constrained from time to time to abandon his most cherished myths. Thus he had abandoned his animism; his Ptolemaic astronomy that assured his position in the center of the universe; his faith in a hereafter that endowed him with eternal life; his belief in the supreme and infinite worth of his person that assured him a position of isolate dignity in an otherwise meaningless and impersonal world; and even perhaps his faith in a God whose attributes, under the impact of man's rationalistic scrutiny, became ever more abstract until He vanished in the metaphysical concept of the Whole. The shedding of these inestimable illusions may be merely stages in his diminishing stature before he himself vanishes from the scene—lost in the icy fixity of his final state in a posthistoric age.[37]

But no present evidence compels us to believe that man's ultimate choice is stasis or catastrophe. Yet if there are preferable alternatives it seems unlikely that they will be discovered wholly by chance, or in disregard of man's basic nature and his environmental necessities. Social response to biological fact ought therefore to be a major concern in a politics of survival. Social understanding of life's realities as revealed by science may influence, perhaps decisively, man's ability to influence his destiny. But a science adequate to this task would necessarily have advanced beyond the science we know. Resolution of the anomalies that now make scientific rationality a dangerous or rejected instrument of policy, qualifying its evident benefits, is a goal toward which advancement would move. A new level in an ascending hierarchy of knowledge has become necessary, at which the integration of the discrete sciences of man and of life can be accomplished.

Notes

1. William Bateson, *Biological Fact and the Structure of Society* (Oxford: Clarendon Press, 1912), 3. Cf. R. J. Halliday, "Biology and Politics: Some Victorian Perspectives," *Journal of Social and Biological Structures* 2 (April 1979): 119–31.
2. Bateson, 34.
3. Bateson, 22.
4. Bateson, 23.
5. Ibid.
6. Bateson, 29.
7. *Biocracy* signifies a political regime in which public policy is consciously or unconsciously guided by knowledge derived from the life sciences including biotechnology. For an early use of the term see Walter B. Cannon, "Biocracy," *Technology Review* 35 (March 1933): 203–27; and a discussion and references in chapter 3, "Health and Homeostasis as Societal Concepts."
8. Lenny Giteck, "A New Beginning, Harry Britt," *The Advocate* Issue 261 (22 February 1979): 12–13; and William Endicott, "San Francisco Homosexual Named to Succeed Supervisor Milk," *Los Angeles Times* (January 1979): 3.
9. John Henry Cardinal Newman, "Knowledge and Faith," cited from *Readings in English Prose of the Nineteenth Century*, ed. Raymond M. Alden (Boston: Houghton Mifflin, 1917), 413.
10. David Easton, *A Framework for Political Analysis* (Englewood Cliffs, New Jersey: Prentice-Hall, 1965), 29.
11. Ibid., 30.
12. Claude Bernard, *An Introduction to the Study of Experimental Medicine*, trans. Henry Copley Green (New York: Dover, 1957), 203. [A republication of the first English translation, 1927.]
13. Note debate on models and modeling following publication of Meadows, et al., *The Limits to Growth*, note 27 below.
14. E.g., the following: James J. Dagenais, *Models of Man: A Phenomenological Critique of Some Paradigms in the Human Sciences* (The Hague: Hijhoff, 1972); Anthony H. Galt and Larry J. Smith *Models and the Study of Social Change* (Cambridge, Massachusetts: Schenkman Publishing Company, 1976); Henry Harris, ed., *Scientific Models and Man* (Oxford: Clarendon Press; New York: Oxford University Press, 1979); Martin Hollis, *Models of Man: Philosophical Thoughts on Social Action* (New York: Cambridge University Press, 1977); Orrin E. Klapp *Models of Social Order* (Palo Alto, California: National Press Books, 1973); and Herbert A. Simon *Models of Man: Social and Rational: Mathematical Essays on Rational Human Behavior in a Social Setting* (New York: Wiley, 1957). For a brief summary description of global models, see *The Global 2000 Report to the President: Entering the Twenty-First Century*, a report prepared by the Council on Environmental Quality and the Department of State, 2 vols. (Washington, D.C.: U.S. Government Printing Office, 1980), 1:43–45 and appendix;

2:24–31. See also Brian P. Bloomfield, *Modelling the World: The Social Constructions of Systems Analysis,* includes bibliography (Oxford: Basil Blackwell, 1986).

15. Note for example, statements attributed to U.S. Secretary of the Interior, James G. Watt, regarding his responsibilities. See Elizabeth Drew, "Secretary Watt," *New Yorker,* 4 May 1981, 104–38; Andy Posztor, "James Watt Tackles Interior Job with Zeal," *Wall Street Journal,* 5 May 1981; and Ron Wolf, "God, James Watt, and the Public Land," *Audubon* 83 (May 1981): 58–65.

16. For cases-in-point see articles published in *Zygon: Journal of Religion and Science,* 1966–. There is, of course, a very large literature relating to science, ethics, and religion.

17. Newman, "Knowledge and Faith," 415.

18. Martin Gardner, *Fads and Fallacies in the Name of Science* (New York: Dover, 1957). Note also Louis Pauwels and Jacques Bergier, *The Morning of the Magicians,* trans. Rollo Meyers (New York: Stein and Day, 1964); and H. L. Nieburg, *In the Name of Science,* 2d ed. (Chicago: Quadrangle, 1970).

19. See Mary Boyce, *A History of Zoroastrianism* (Netherlands: Brill, 1975); Miles M. Dawson, *The Ethical Religion of Zoroaster* (New York: AMS Press, 1969); and C. F. Potter, *The Story of Religion as Told in the Lives of Its Leaders* (Garden City, New York: Garden City Publishing Co., 1929), 77–99.

20. Cf. Miguel de Unamuno y Jugo, *The Tragic Sense of Life in Men and Peoples,* trans. J. E. Crawford Flitch (London: Macmillan, 1926).

21. Cited from Thomas Bethell in "Burning Darwin to Save Marx," *Harper's Magazine* 257 (December 1978): 31–38 and 91–92.

22. Among biologists note, for example, the examination of ethical concepts and values by Rene Dubos, Theodosius Dobzhansky, Alfred Emerson, Hudson Hoagland, Van Rensselaer Potter, and Conrad Waddington, among others.

23. Note Theodore Roszak, *The Making of a Counter Culture* (Garden City, New York: Doubleday, 1969), and *Where the Wasteland Ends: Politics and Transcendence in Postindustrial Society* (Garden City, New York: Doubleday, 1972). See also Ernest Callenbach, *Ecotopia* (Berkeley, California: Banyan Tree Books, 1975) for description of a fictional counterculture society.

24. John A. Hostetler, *Amish Society* (Baltimore: Johns Hopkins Press, 1963); and William I. Schrieber, *Our Amish Neighbors* (Chicago: University of Chicago Press, 1962).

25. *Ecologist* 2 (January 1972): whole issue. Published in book format as *Blueprint for Survival: By the Editors of the Ecologist,* ed. Edward Goldsmith, et al. (Boston: Houghton Mifflin, 1972).

26. Ibid., 8.

27. E.g., Donnella H. Meadows, et al., *The Limits to Growth: A Report for the Club of Rome's Project on the Predicament of Mankind* (New York: Universe Books, 1972); Mihajlo Mesarovic and Edward Pestel, *Mankind at the Turning Point: The Second Report to the Club of Rome* (New York: Dutton, 1974).

28. H. S. D. Cole, et al., *Models of Doom: A Critique of "The Limits to Growth"* (New York: Universe Books, 1973); note also Eric Ashby, *A Second Look at Doom,* the twenty-first Fawley Foundation lecture (Southampton, England: University of Southampton, 1973).

29. Jan Tinbergen, et al., *Reshaping the International Order: A Report to the Club of Rome* (New York: Dutton, 1976).

30. E.g., see David Ehrenfeld, *The Arrogance of Humanism* (New York: Oxford, 1978).

31. The literature on the feasibility of life and human society in extraterrestrial space is now extensive. As a sample see Karl Ernst Shaefer, ed., *Bioastronautics* [especially chapter 10, "Space Ecological Systems"] (New York: Macmillan, 1964); and *Extraterrestrial Life: An Anthology and Bibliography,* comp. Elie A. Shneour and Eric A. Ottesen (Washington, D.C.: National Academy of Sciences, Natural Resource Council, 1966).

32. Gerard O'Neill, *The High Frontier: Human Colonies in Space* (New York: Morrow, 1977); and "Colonization of Space," *Physics Today* 27 (September 1974): 32–40.

33. *Newsweek,* 27 November 1978, 100.

34. Catastrophe theory as a specialized field of mathematics has interesting behavioral and (possible) ecological implications. See Dixon D. Jones, *The Application of Catastrophe Theory to Ecological Systems,* [Research Report RR-75-15] (Laxenburg, Austria: International Institute for Applied Systems Analysis, 1975); and Tim Poston and Ian Stewart, *Catastrophe Theory and Its Applications* (London/San Francisco: Pitman, 1978).

35. Roderick Seidenberg, *Posthistoric Man: An Inquiry* (Chapel Hill, North Carolina: University of North Carolina Press, 1950).

36. Ibid., 180.

37. Seidenberg, *Posthistoric Man,* 237–38.

3
Health and Homeostasis as Societal Concepts

Among the things about himself that man has understood poorly are the effects of his behavior on the prospects for his continuing existence. If humans fail to find answers to the problem of their behavioral-ecological relationships and to be guided by this knowledge, their future may be too brief to find answers to most of the other perplexing questions about human behavior. Only recently have humans found means to discover the true nature of their ecological situation; and only recently has it become imperative to assess these findings correctly and to act upon their implications while there is yet time for choice. For human society is now at one of the critical stages in its evolution, and whether it can surmount it and move to a "higher" level of existence is conjectural.

From the time that humans discovered technology and began the unprecedented multiplication of their numbers, arrival at this decisive stage was probably inevitable. Only cosmic intervention could have prevented it. This is the stage at which the pressure of human demands upon the earth stresses its capacity for sustained yield and self-renewal. This stage is not reached at the same time for all resources or elements in the planetary life-support system, nor is it reached everywhere simultaneously. It marks, for any society, the beginning of a phase in its history when increases in numbers of people or in their demands upon their environment becomes progressively disadvantageous to the prospects for their survival.

The danger of continued expansive growth, both to individuals and to society generally, lies in its progression to a point at which further expansion and even renewal becomes impossible. If an essential element in the life-support system were to be destroyed, (e.g., the atmospheric ozone layer) dependent organisms—including man himself—could be endangered. Short of this point, in order to survive—certainly to prosper—a society would have to stabilize its relationship to its environment. The consequence would be societal homeostasis—a society whose dynamism would be internalized rather than directed toward expansive growth. Whether civilized technological man is capable of achieving such a society, under what conditions, and at what costs, is perhaps the most important question to which a science of human ecology could be directed. In 1912 an English social theorist, Edward Isaacson, prefaced his book *The Malthusian Limit* with the statement that the problem of how human society could limit its numbers "to figures which can make best use of the world's natural resources" was "The Most Important Question for the Human Race."[1] More than half a century later, this may still be the most important question.

This is at least the fundamental biopolitical question, but we here can do no more than to indicate some of the attempts to answer this question and to suggest the directions that research now appears to find most promising. Basic to an understanding of social evolution, or of the prospects for its rational control, is a prior understanding of the way in which the forces that make for diversity, stability, and growth in society influence the course of its development. We have as yet no adequate theory explaining the interactions of these forces in relation to the integrity of human society. The closest approaches have been formulated around such "organizing concepts" as social health, homeostasis, and less directly in the Gaia hypothesis.[2] For this reason our focus is upon the feasibility of these concepts as approaches toward understanding the effect upon the human ecosystem of those conditions that we describe as *diversity* and *stability* and which appear to be unsustainable under conditions of uncontrolled expansive growth.

The Provenance of Homeostasis

The concept of societal homeostasis, although developed in part by scientists, is not in any rigorous sense a scientific theory. Among the categories of intellectual formulations it is located between conjecture and hypotheses. Its history is long, but discontinuous. It has attracted the interest of individual scholars, but has not induced a major systematic intellectual effort continuously through several successive generations of scholars. Yet there are indications that such an effort may be emerging, as the necessity for human society to adjust its behavior to the finite realities of its physiological and environmental parameters becomes increasingly evident.

The following definition was adopted in 1964 by the Eighteenth Symposium of the Society for Experimental Biology:

> Homeostasis in its widest context includes the coordinated physiological processes which maintain most of the steady states in organisms. Similar

> general principles may apply to the establishment, regulation, and control of steady states for other levels of organization. It must be emphasized that homeostasis does not necessarily imply a lack of change, because the "steady states" to which the regulatory mechanisms are directed may shift with time. But throughout the change they remain under more or less close control.
>
> Such a concept can therefore be applied to organizations at cellular, organ system, individual, and social levels. It may be considered in relation to time intervals ranging from milliseconds to millions of years. Its essential feature is the interplay of factors which tend to maintain a given state at a given time.[3]

The concept of dynamic homeostasis is implicit in the following observation by biologist Rene Dubos:

> There is evident everywhere in nature a close correspondence between most of the structures and activities of living things and the needs imposed upon them by their places in nature. The necessary outcome of this relationship is a dynamic equilibrium which implies inseparability of life and environment, and which has fitness as a consequence.[4]

The principal intellectual problem in the recent history of thought on societal homeostasis has been to ascertain the validity of extrapolating a physiological theory to a social hypothesis. Societal homeostasis as a philosophic proposition is as old as articulated concepts of man's relationship to nature; as a scientific proposition, however, its origin appears to be in experimental physiology, and in this sense it is a "modern" concept. For the greater part of the century following publication of Claude Bernard's *An Introduction to the Study of Experimental Medicine* (1865),[5] the principal contributions to thought on homeostasis were made by biologists. Two lines of inquiry developed out of Bernard's concern with organic interactions in relation to environmental factors. The first of these was physiological homeostasis and is exemplified by Lawrence J. Henderson's *The Fitness of the Environment* (1913),[6] and especially by Walter B. Cannon's, *The Wisdom of the Body* (1932).[7]

This general line of inquiry drew upon the physicochemical sciences and biology; its focus was upon the conditions conducive to the existence and maintenance of organic life. Henderson's concern was primarily with the physicochemical parameters of life, with Bernard's *milieu extérieur.* Cannon was initially concerned with the interactions and adjustments through which an organism maintains its functions in relation to the impact of its external environment, with Bernard's *milieu intérieur.* Growing out of these related areas of investigation, an ecological concept of society has emerged that is now beginning to influence public policy, to redirect and reinforce scientific interest in the larger aspects of the evolution of society and especially in the mechanisms of social control.

A different but not necessarily incompatible approach to the maintenance of living systems has been James E. Lovelock's Gaia hypothesis.[8] The name Gaia (Greek goddess of the Earth) was taken to symbolize the total planetary system of the earth which appears to exhibit the behavior of a single organism. Contrary to the theory that life exists on Earth only because the physical conditions have been right for its emergence, Lovelock advanced the hypothesis that life actually defines the material conditions needed for its survival and responds, through homeostatic and compensatory reactions, to maintain those conditions. This concept appears to have been a factor in stimulating the International Council of Scientific Unions to propose a new large program of research—"The International Geosphere-Biosphere Program: Global Change."[9] Thus we see the growth, elaboration, and correction of scientific concepts, that in the long-run may give humans a new view of their situation on Earth.

The derivative line of inquiry, on what may be called societal or social homeostasis, was initiated, among others, by Cannon himself in the concluding chapter of *The Wisdom of the Body* and was developed in a paper entitled "Biocracy: Does the Human Body Contain the Secret of Economic Stabilization?" (1933)[10] and in his presidential address, "The Body Physiologic and the Body Politic," before the American Association for the Advancement of Science (1941).[11] Cannon's purpose was to stimulate inquiry into the possibilities of relationships between organic and organizational behavior. He was interested in ascertaining the validity of apparent analogies between the functional organization and behavior of physiological and social systems. The response to his concepts of homeostasis, both physiological and social, was mixed. But arguments over physiological homeostasis were primarily directed to the limitations of the theory and to the extent of its applicability.[12] There was general acceptance of its basic reasoning, which could be demonstrated experimentally in physical systems.

Homeostasis as a social concept was not only more difficult to demonstrate by reference to empirical events, it was also more difficult to formulate in theoretical terms. This latter difficulty was complicated by the presence of older concepts of social evolution in which the condition of homeostasis was implicit. The case has been one of "complication by association." Traditional folk-wisdom and a large number of social philosophies have postulated the actuality, necessity, or desirability of a "moving equilibrium" relationship between man and nature. With the growth of scientific knowledge, analogies from the sciences have been sought to test or to confirm various theses regarding social stability. A brief reconnaissance of some of these propositions may help to place the present problem in perspective.

Historical antecedents of homeostasis theories often differ in emphasis from our contemporary formulations of societal homeostasis. Yet there is very little in man's effort to understand his relationship with nature that does not become pertinent to a consideration of public action to guide or control those relationships. Control of environmental relationships requires control of some sort over the behavior of people. What people have thought or felt regarding these relationships is therefore pertinent to public action. Even self-control of behavior needs social reinforcement. Scientific knowledge is not automatically persuasive, and it is therefore necessary to understand the context within which new concepts must find acceptance if one wishes to promote consideration of these concepts in society.

Modern thought on man-nature or man-environment relationships can be identified at least as early as 1542 with the publication of Jean Fernel's *De Abditis Rerum Causis* (On Hidden Causes).[13] Fernel's study was modern in the sense that it approached the causes of disease and the relationship of man to nature from a point of view that can be called "scientific." Fernel's approach to his subject has a holistic or ecological orientation that places it in the main current of thought leading toward contemporary ecological and systems theory (cf. Sherrington.[14]) Also relevant to thought regarding the reactive behavior of living systems to environmental effects is the proposition of Thomas Kuhn[15] that revolutions in scientific thought are reactive to the total intellectual development of society. From these examples one can observe in a general way the process by which the concept of societal homeostasis gradually took form as a significant current in social thought. It appears first by implication in philosophical observations concerning man and nature, becomes explicit in eighteenth and nineteenth century theories of social mechanics and self-regulating economics, and in the twentieth century becomes a subject of inquiry upon which several sciences converge, and which grows in importance as science is increasingly invoked as a rationale for social control.

Complications in the course of thought on societal homeostasis have appeared when that thought passed from description to prescription. Three examples may suffice to illustrate the effect of the mixing of science and social policy on the concept of societal homeostasis. From among a large number of possibilities, the social consequences of the theoretical propositions of the Physiocrats, of Thomas Malthus, and of Herbert Spencer illustrate the difficulty of predicting or controlling the outcomes of efforts to apply science-based knowledge to social policy questions. The principal focus of each of these three sources of social theory was upon political economy, and the social theses associated with them were essentially integrative or synthesizing. In each case, biology, sociology, economics, and ethics were intermixed; all of the theses predating the phenomenal expansion, specialization, and segregation of the science occurring after 1900.

Physiocracy

The Physiocrats were a school of economists, largely French, whose direct influence was confined to the eighteenth century and among whose principal advocates were François Quesnay (1694–1774), Victor Riqueti de Mirabeau (1715–1789), Pierre Samuel du Pont de Nemours (1739–1817), and the Baron d'Holbach (1723–1789). The basic thesis of the Physiocrats was that the actions of man through government would not be beneficial unless consistent with the laws of nature. The political economy of Physiocracy relied for the production of wealth upon the operations of natural systems undistorted by the manipulations of governments and other "artificial" institutions. This attitude led to a fixation upon certain points of doctrine which, however, did not necessarily follow from the basic thesis. Principal among these doctrines was the belief that agriculture was the primary source of all wealth and that a single tax on land (*impôt unique*) was the most equitable and least harmful form of taxation. Laissez-faire economics became a logical consequence, but not the only possible consequence, of the doctrine. Laissez-faire made sense if it were true that natural forces, including those physically engineered by man, operated to maintain a total equilibrium in society. If human society functioned within its environment as a self-regulating, self-renewing system, the interposition of man through public authority could hardly improve upon nature and would more likely create obstructions that would be harmful to stability of the system and hence to the general welfare.

As with any concept or intellectual movement in history, it is difficult to appraise precisely the influence of the Physiocrats on the development of a science of society. Did their writings and their influence contribute to the evolution of homeostasis as a social concept or did their dogmatic premises induce indifferent reactions? Was their socioeconomic analysis totally invalid or was it, although misconceived, suggestive of productive hypotheses? Is it fanciful to conjecture that their theories of political economy might be important primarily as a culture dish in which the germ of the idea of socioecological homeostasis was nurtured? In a manner analogous to the history of penicillin, could it be that an element in their theories long known and not regarded as the main focus of their concern might subsequently be found to have a new significance and a revised place in the history of social thought?

The relative importance of the components of a set of ideas may change over time. The laissez-faire arguments of the Physiocrats were congenial to the commercial and manufacturing classes of the nineteenth century—not so their idea of the higher efficiency of natural systems (except to the extent that the market economy as they conceived it could be called a natural system). Physiocracy as a general orientation of political and economic thought could not have been a workable social hypothesis for the business enterprise system or the technology or politics of the nineteenth and early twentieth centuries. It was a philosophy better adapted to a view of the world as a closed system than to the prevailing view of unlimited horizons, indefinite expansion, and the confidence that man could improve over all aspects of nature.

A collateral element in the political theory of Physiocracy was its concomitant system of natural ethics. Here one encounters another current in the history of social thought: the concept of natural theology and natural law, an area of philosophic concern as old as philosophy itself and exemplified in the previously cited work of Jean Fernel. Among the Physiocrats, it was the Baron d'Holbach who summarized most completely the ethical assumptions of the school in his *Éthocratie, ou le Government Fondé sur la Morale* (1776).[16] The political theory of ethocracy could be dismissed as an intellectual relic of no more than antiquarian interest except for the persistence of the effort to fuse science, ethics, and politics into a coherent, operational, social philosophy. Viewed retrospectively the biocratic concepts of Walter B. Cannon and the biology-based ethics of Alfred Emerson, presently to be considered, are in a tradition that gains pertinence (although as yet not popularity) as the advancement of science gives rise to ethical problems for which traditional ethics do not seem to provide adequate answers. The point to be made here is that these concepts contributed to a body of thought about the "health" of society and that they assumed an interactive relationship between man and nature that when properly maintained had a socioecologically stabilizing effect.

Malthusianism

The influence of Thomas Robert Malthus (1776–1834) on attitudes toward social stability is too well known to require more than the briefest comment. His most relevant work, *An Essay on the Principle of Population as It Affects the Future Improvement of Society* (1798 and revisions to 1826), aroused a storm of protest and continues to be a subject of controversy.[17] In brief, Malthus postulated the ultimate control of the growth of human populations by the interacting checks of war, famine, pestilence, vice, and misery. He criticized the society of his day for exaggerating the expansive tendency of human populations through doles to the poor and bounties for large families. He had little confidence in the capacity of society to establish a rational system of self-regulation. Societal homeostasis as conceptualized by Malthus was attained at costs which were repugnant to liberal and humanitarian thought. But operation of the Malthusian thesis was suspended, at least for Europe, by the opening of new lands in America, Australia, and South Africa and by phenomenal developments in production technology. Yet in the absence of demonstrably valid alternatives to Malthusian controls, their *ultimate* force could no more be refuted than proved.

As is often the case with apparently inexorable and disagreeable eventualities, Malthusian homeostasis has been repeatedly rejected without being absolutely disproved. Nearly every contemporary popular theory of government and society, from traditional conservatism to Leninist Communism, regards Malthusianism as a false and harmful dogma. *"Malthusianisme, Voilà L'Ennemi,"* declared a headline in the French journal *Combat* in 1950, on the same page that carried a series of articles, *"Les sciences de la vie,"* by Alain Sergent[18] that could be read to an opposite conclusion. Malthusianism was criticized as a "politics of prudence," restrictive of economic growth. These and other indictments of Malthusianism can be found in publications as diverse in outlook as those of the Soviet Union, the American academic and business communities, and the Vatican. Yet the political unpopularity of the Malthusian thesis has not discouraged demographic studies; its stark prognosis may have stimulated them. It is also conceivable that high-minded dislike of the conclusions reached by Malthus and equal mistrust of the implied alternatives of socially enforced control have tended to channel demographic studies toward descriptive rather than prescriptive goals. Nevertheless policy-oriented population studies have been increasing in prominence and probably in numbers (cf. Hardin[19] and Berelson[20]). But comprehensive estimates of the socioecological consequences of alternative methods of population control have not been adequately explored, nor have adequate analyses been done on the political conditions under which alternative methods could be made effective. It is probable that studies of these possibilities are under way; it is highly desirable that they be undertaken as rapidly as the state of knowledge permits. Also of importance is recognition that the socioecological aspects of population dynamics include the movements of peoples, their migration and concentration. Immigration can alter the gene pool and ethnic character of a region as effectively as indigenous birth rates.

Spencerian Social Theory

The influence of Herbert Spencer (1820–1903) has declined sharply since the beginning of the twentieth century. His attempt to create a synthesis of scientific knowledge, to develop a coherent unified theory of society, has not been congenial to the attitudes and values of the increasingly specialized sciences or compatible with the trends of democratic and socialist political thought. His synthesis suffered from the appearance of an effort to force facts into conformity with a logical system. Many of his conclusions have been discounted by subsequent evidence. Nevertheless his *Synthetic Philosophy*[21] represents a vast and erudite intellectual effort, and contains analyses of social behavior and development that might repay further investigation.

Spencer was greatly concerned with variety and stability in organisms and organizations. He believed that the growth of individuation and of variety were forces for stability. In this he followed theses earlier advanced by Spinoza and Leibniz and may have anticipated in his "law of the persistence of force" and "law of the multiplication of effects" the stabilizing influence of variety in complex systems. His concern with the simultaneous processes of differentiation and integration and the functional values of variety have since been paralleled in a more closely defined manner in the study of ecology and cybernetics. Spencer's concept of "evolution under all its aspects" was that of " . . . progress towards a moving (dynamic) equilibrium completely adjusted to environing actions."[22]

In reviewing Spencer today, it is not always easy to discriminate his interpretation of what he thought was actually happening from what he thought was probable or desirable. His phrase "the survival of the fittest" aroused hostility among those who found it easier to place their own interpretation on Spencer's choice of words than to attempt to discover what the phrase actually meant to him. Its intended meaning could not be understood outside the context of Spencer's evolutionary hypothesis. But this hypothesis was no more acceptable to believers in indeterminant growth than the survival of the fittest was to those who saw it as an affront to Judeo-Christian ethics. Failing to foresee the explosions of population and technology in the twentieth century, Spencer believed that the tendency of human society, as of all organic life, was toward equilibrium with its environment. In the concluding paragraph of his *Principles of Biology*[23] he wrote:

> In approaching an equilibrium between his nature and the ever-varying circumstances of his inorganic environment, and in approaching an equilibrium between his nature and all the requirements of the social state,

> Man is at the same time approaching that lowest limit of fertility at which the equilibrium of population is maintained by the addition of as many infants as there are subtractions by death in old age. But in a universe of which all parts are in motion and every part is consequently subject to change of conditions, neither this equilibrium nor any other equilibrium can become complete.

Comparisons of Equilibrium Theories

Different in origin, focus, and emphasis, the foregoing precedents to a concept of societal homeostasis possess several significant characteristics in common. All of them attempted to construct a theory of social behavior on a foundation of science as they perceived it. In this they were not unique. But, unlike Marx, for example, they believed that physical and biological forces in the human environment were more significant than human institutions in limiting or guiding the course of history. Because they assumed an ultimate necessity for society to accommodate its behavior to "natural laws" and to achieve a self-renewing steady state, they were, in a fundamental sense, "conservative." This conservatism did not necessarily extend to the dominant values or institutions of their times—they were often subversive of prevailing assumptions. All of them implied ethical conclusions which were often at variance with religious or political orthodoxy. Their common concern with stability did not always extend to variety. Only to Spencer was the process of individuation an indispensable element of the total thesis. Yet none of the theories were inimical to diversity and, at least by implication, variety in human society was in every case concomitant to the social condition of homeostasis that would prevail if the admonitions of the theorists were heeded.

The usefulness of these theories to the study of societal homeostasis and to the effects of diversity and stability in human societies obviously does not rest (at least at present) on demonstrable validity. Yet they possess certain values that it would be "unscientific" to neglect. Among these are potential contributions to social perspective and to insight. Through them one may better understand how men have perceived their ecological relationships. They may also suggest or document the existence of deep-seated physiological problems with which man must cope in his effort to understand and to "dominate" the natural world. The capacity of a culture, at any stage in its history, to accept new attitudes is dependent on the way in which its assumptions and institutions have developed prior to that stage. Social change does not proceed from nowhere. Science likewise proceeds from where it has been, and benefits as often from knowledge of past errors as from propositions capable of development into future theories. Men of demonstrated intellectual capability have sought diligently to understand the conditions, mechanisms, and consequences of social stability and instability. It would be improvident to disregard the possible benefits of their thought.

The Wisdom of Society?

Are there innate properties in societies that normally protect them from oscillations so excessive as to jeopardize their survival? This is a question of general bioecology, because there are many aggregations of organisms that maintain relatively stable relationships over extended periods of time. These aggregations include both ecosystems and societies; and there are many forms of societies other than those that are distinctively human. It is plausible to argue that self-maintaining systems of living organisms are inherently homeostatic. Walter B. Cannon wrote that:

> Hardly any tendency in a nation continues to the stage of disaster; before that extreme is reached corrective forces arise which check the tendency and they commonly prevail to such an excessive degree as themselves to cause a reaction. A study of the nature of these social swings and their reversal might lead to . . . means of more narrowly limiting the disturbances. . . .
>
> . . . It would appear that civilized society has some of the requirements for achieving homeostasis but that it lacks others, and because lacking them it suffers from serious and avoidable afflictions.[24]

It is not clear whether historical or anthropological evidence supports Cannon's thesis. Various societies have, in effect, been self-destroyed. We have little reliable knowledge concerning self-maintaining institutions in society or whether there is a residual wisdom of society that continues to protect the species, analogous to the automatic protective responses that constitute the "wisdom of the body." Given this uncertainty, would it be meaningful to apply the concept "health" to those conditions of society which appear to be homeostatically maintained? Does the concept of social health, widely and loosely used in social polemics, also have a meaning comprehensible to science? An examination of contemporary thought on social health and homeostasis may indicate sources from which answers to these questions may possibly be derived. Whether or how far the answers may be valid is another level of inquiry.

Convergent Concepts

The historical development of the concept of societal homeostasis was largely an extension of physiological and natural-order propositions to theories of society. Political economy, as for example expressed in Physiocracy and social Darwinism, was the medium through which the concept of societal homeostasis was most often formulated. But by midtwentieth century, several new lines of science-related reasoning had converged to reinforce the concept and the interest of some scientists and in its further exploration. Its focus was no longer primarily on political or economic behavior, although the most explicit extension of homeostasis concepts to politics that has thus far occurred was formulated by Walter B. Cannon in his theory of biocracy. The mainstream of interest in societal homeostasis in the recent past has been in the study of sociology, and more specifically in theories of organization and of human ecology. The prescriptive, normative, philosophical treatment of the concept of societal homeostasis has largely been replaced by an analytic approach. This shift is in part a consequence of the maturing of the sociobehavioral sciences, and in part a result of the incompatibility of the philosophic concept of homeostasis with the political and economic doctrines that increasingly dominated industrial and democratic societies particularly after 1865.

There is now, however, a third reason for the growth of interest in understanding how and to what extent homeostasis mechanisms are operative in society. This is the practical necessity for dealing with homeostatic responses in the course of applying cybernetic models to the study of social and organizational behavior.[25] Following World War II, the sharpening of a number of fundamental concepts in the physical sciences prepared the way for their extension into the biobehavioral and social sciences. The convergence of concepts from mathematics, several sciences, and several fields of engineering has been involved in this process of theory building and it is at present impractical to attempt to identify the order and degree of convergent or reciprocal influences. But it is clear that information theory, cybernetics, and general systems thinking have added greatly to the organization of knowledge on homeostasis in organisms and organizations.

It is not yet clear whether a comprehensive integrative theory of society is emerging from this complex diversity of thought. It is, however, evident that a foundation for social theory is being laid down that is much broader and firmer than we have had before. But before examining this foundation it will be useful to look at the most explicit formulation of the concept of societal homeostasis—the biocratic theory of social control proposed by Walter B. Cannon.

Biocracy—The Politics of Homeostasis

Outcomes in science have often been apprehended intuitively before the marshalling and analysis of evidence and the construction and testing of theory have been consummated. The naturalistic and organismic theories of society may possibly be examples of this prescience. But their sources have more often been in philosophy than in science. Not until Bernard and Darwin had done their work, was there a conceptual foundation adequate to support a theory of societal homeostasis that could be referenced to empirical and demonstrable evidence.

A singular contribution toward a biologically inspired social theory was offered by Walter B. Cannon in the early and rapidly worsening stage of worldwide economic depression. His views were initially published in the epilogue to his book *The Wisdom of the Body*[26] and in an article entitled "Biocracy."[27] He asked, "Does the human body contain the secret of economic stabilization?" Assuming that some minimal degree of stability was necessary for the health or survival of both organisms and organizations, Cannon suggested the application to society of the principles of self-regulating mechanisms and of redundant capacity. He advocated the provision of indicators or early warnings of disturbance to stable states in society—a proposition partially embodied after 1946 in the Council of Economic Advisors of the United States and proposed in a national system of social indicators,[28] in provisions for the monitoring, surveillance, and interpretation of socioeconomic change.[29] If stabilizing mechanisms are not innate, but must be invoked, timely indication of the need for them is necessary. Proposals to establish a forecasting capability in the United States government were introduced into the ninety-ninth and one hundreth Congresses by Senator Hatfield of Oregon but had not received hearings at the time this book was completed.

Cannon argued for the priority of stability over economy in social processes, but also for specially organized controls to maintain the constancy of economic processes. "In the light of biological experience," he declared, " . . . social stabilization should be sought, not in a fixed and rigid social system, but in such adaptable industrial and commercial functions as to assure continuous supplies of elementary human needs."[30] An outcome of the homeostatic society in Cannon's view should be a shift of effort and emphasis from the routine or operational details of sustaining the economy to the higher goals of civilization. Physiological homeostasis frees those activities of the brain associated with conscious life from the necessity of unremitting conscious attention to the maintenance and protection of the body as a living system. Because of this automatic responsiveness of the human organism to its environment the human personality is "free to enter into agreeable relations with our fellows, free to enjoy beautiful things, free to explore and understand the wonders of the world about us, develop new ideas and interests, and to work and play untrammeled by anxieties concerning our bodily affairs."[31] Societal homeostasis would extend this principle of automatic systems maintenance to the economy, thus freeing human effort for advancement of civilization in its intellectual, aesthetic, and ethical aspects.

In 1940 Cannon reiterated this physiological view of social organization in his presidential address before the American Association for the Advancement of Science (AAAS).[32] In this address he gave more attention to the political implications of his thesis. His views of societal homeostasis extended not only to societies, but to the behavior of nations. They were recalled in a tribute to Cannon offered by academician C. S. Koshtoyants[33] of the Institute of Experimental Biology of the U.S.S.R. when, in 1960, at an international symposium on homeostasis (dealing almost exclusively with physiological homeostasis) he declared that: "We shall remain faithful to the words of Cannon and Pavlov and shall fight for peace, for genuine human homeostasis, for the happiness of all men on our planet."

An obvious break in the analogy between the body physiologic and the body politic was the absence in society of a control center analogous to the brain. Society as a whole has no consciousness—no mind. But Cannon distinguished the creative functions of the brain from its role as an automatic center for homeostatic control. In the social body he identified " . . . the revealers, the creators, the directors of affairs . . ." as simulating the higher functions of the brain. He was less clear in his identification of automatic control mechanisms that would free society for more creative functions. He came no closer to proposing a coordinative system than to call for ". . . a thorough cooperation of functional groups. . . . "

Cannon was thus confronted by the same enigma that Herbert Spencer encountered in his effort to conceptualize a society as a moving equilibrium in which variety and integration increased automatically. The enigma was complex because it simultaneously not only posed the question of what forces controlled social behavior, but involved in juxtaposition considerations of what was, what might be, and what should be. The question of whether there was a homeostatic governance of society could not be resolved by any available methodology. Malthusian controls could not be disproved; they could be postponed or evaded, but could not be reconciled with human health or happiness. Spencer believed that there were indeed automatic tendencies in society that would maintain social "progress" and equilibrium if undisturbed by unwise human interference. Cannon conversely hypothesized that homeostatic controls might be instituted to protect and advance human welfare, but that their institution would require rational human action.

Cannon was therefore ultimately forced to confront the problems of how political decisions are made and how political leadership is selected and replaced in a homeostatic society. For this task he lacked both method and theory. He could not make his concept operational. His essay on "Biocracy" ended on a pessimistic note:

> I do not look forward with hope, however, to a real "biocracy;" and "technocracy" also is not likely soon to have the influence which its promoters desire. For years to come the technocrats and the biocrats will "view with alarm" the events which occur about them. And it will be a long time, I fear, before they will be able to "point with pride" to the value of their suggestions. The technocrats and the biocrats must wait, because the democrats have the votes![34]

Cannon, be it noted, was advancing the outline of a hypothesis by way of suggesting an area for extended inquiry. He did not present a developed theory although his concept was explicit and its essential assumptions were clear. It is also evident that he believed that biocracy—an incipient theory of natural government—could provide a sounder sociopolitical system than the form of popular democracy then prevailing. Implicit in his thesis was a concomitant concept of social health. At the conclusion of his AAAS address[35] he declared that: " . . . insofar as social stability fosters security, both physical and mental, of the members of a social organization, so also it fosters their higher freedom, assuring them opportunity for self-support, replacing fear by confidence and providing for health and responsible leisure."

A decade later, a French journalist, Alain Sergent, described the social role of biology in similar terms:

> La "biocratie," c'est-à-dire l'organisation selon les lois fondamentales de la vie, représente notre seule chance de salut à un moment de notre évolution où les métaphysiques et les systèmes, résidus de civilisations archaiques, empoisonnent encore la vie des hommes. . . . [36]

Available evidence does not indicate whether Sergent owed his biocratic concepts to Cannon, yet like Cannon, he saw *biocratie* as a means of restoring social equilibrium and preventing "the monstrous errors daily committed" in industrial society. A new social science based on biology would be able to provide an informed conscience and a norm by which valid judgments could be made concerning social issues.

A more elaborate application of biological concepts to social theory was essayed by zoologist Alfred E. Emerson[37] in a paper entitled "Dynamic Homeostasis: A Unifying Principle in Organic, Social, and Ethical Evolution" (1954). An adequate critique of this paper would exceed its length and cannot be undertaken here. Emerson's attempt to base a system of ethics upon the concept of societal homeostasis was criticized by Patrick Romanell (1955),[38] a philosopher who argued that: "No social principle regarding facts alone (such as social homeostasis) can ever resolve value problems as such, since its self-appointed task is to interpret *existing* social conditions, not *ideal* ones, which by their very motive as ideal must always be postulated as better than actual conditions at any given period of time."[39] To a considerable extent, the reasonings of Emerson and of Romanell appear to have bypassed one another. Emerson's effort to build a bridge between biology and ethics may have fallen short of success—less through the inadequacy of his conceptual design than through the insufficience of his building materials—through the insufficience of knowledge concerning the causes and effects of human behavior. And the argument was confused by an unduly narrow construction that Romanell placed upon the attitude toward ethics which he read as implicit in the homeostasis concepts of Cannon and Emerson. "The homeostatic theory of ethics, inspired directly in Cannon (indirectly in Darwin)," he declared, "may be defined as the doctrine . . . that the rules of prudence are the only reliable guides to life."

Whatever the merits of the arguments of Emerson and Romanell, it is nevertheless clear that the concept of societal homeostasis is inevitably involved with concepts of health and of ethics. As Romanell observed, " . . . problems of health involve ultimate issues of an ethical nature." The complex interrelatedness of the issues and the differing modes of reasoning involved in synthesizing biological, behavioral, social, and ethical elements into an integrated theory of social evolution make both the effort and its critical examination an exceedingly difficult intellectual task for which competence is required not only in one, but in all four of the component fields of inquiry. The task would seem to call for the audacity of Renaissance genius equipped with twenty-first century information. No one since Herbert Spencer has seriously attempted the task, but possibly, like the ascent of Mt. Everest, the effort must be made many times until the mapping of the intellectual terrain and the technology of synthesis one day permit a Darwin or Wallace of the social sciences to reach the summit.

Toward a Science of Social Systems

If there is a "wisdom of society," its present and continuing reality has not been demonstrated. It remains, like social health, largely a metaphysical concept. As scientific concepts, social health and homeostasis do have empirical referents; both are plausible in the sense that the conditions that they represent have been described at lower levels of biological integration. Their confirmation or rejection at societal levels of integration is contingent on the development of a biobehavioral or socioecological science adequate to the task. There are indications that a more adequate science *is* developing. It is also evident, at least to some observers, that this development is not occurring fast enough. This is not an age of utopias and we do not have images of future states that appear to be consistent with human needs and tolerances, although we are well supplied with conjectures regarding the future. Economist Kenneth E. Boulding has observed:

> As we move towards the discovery of "truer" knowledge of social systems—by which I simply mean images of social systems with smaller and smaller ranges of system-maintaining surprise—the success of actual social systems in the ecological successions of human history will depend more and more upon the extent to which they are based on these more "sophisticated" images, and on the complex systems of information-collection and analysis which support them.[40]

Social Stability as Systems Behavior

Insufficiency of resources is not a principal reason for the failure of science to be of much assistance in addressing the problems of an off-balance society. The generally unsatisfactory results of attempts to extrapolate biological behavior to social behavior have been explained as consequences of insufficient attention to the fundamental differences in the forces for integration at biological and social levels. Throughout the long history of organismic theories of society, the method of formulation has been "reasoning by analogy." Prior to the emergence of an empirical social science, the critics of organismic theories were no better equipped than the advocates to examine the validity of the concepts. The field of argument was essentially philosophical.

Social and behavioral science after 1900 progressively undertook the exposure of naive extrapolations. It was capable of dissecting unsophisticated analogies and of cutting out much of the evidence that had been alleged to support a biology-based sociology. Distrust of biological analogies became endemic among social scientists. But unfortunately, justifiable skepticism of biosocial analogy often tended to grow into a dogmatic rejection of the relevance of biological principles to social behavior. This rejection may have been prompted by more than scientific rigor. It appears to have been motivated also by an emotional rejection of the values that seemed implicit in Darwinian concepts. As Sir Charles Sherrington[41] observed: "Scientific outlook is always part of its age."

The social science of the first half of the twentieth century was heavily infused by concern with the social consequences of the industrial revolution, many of which were destructive to humane values and social integration. The principles of organismic and naturalistic theories of social control offered no apparent protection from the abuses of uncontrolled industrialism. The historical determinism of Marx was more congenial to the age than the biological determinism of Malthus. Nineteenth century man had substituted technology for God and nature as shapers of human society and its environment. If there were a homeostatic force in nature that, left untrammeled, would maintain society in a "healthful" equilibrium, its benefits could not be realized merely through avoiding the mischievous interposition of political institutions. Paradoxically, the interference of progressive technology with nature must also be prevented, and to do this would almost certainly require the regulatory force of government. Upon this pragmatic rock of political necessity modern theories of self-guiding social systems have foundered.

If there were a homeostatic wisdom inherent in human sociality, it would have been the consequence of natural selection, the result of an agelong evolutionary process. But would this conjectured response, internalized in hominid society, continue to be operative under conditions of civilization, advanced technologies, and complex social systems? If operative today, it seems apparent that the effects of such internalized behavior patterns would be unlike those obtaining under prehistoric conditions. Like the flight and fight responses that produce internalized stress in civilized man, a residual homeostatic impulse in society, if it existed under present-day conditions, could conceivably give rise to a pervasive, indefinable, social stress or malaise.

Cannon believed that there was evidence that homeostatic mechanisms were in part and imperfectly operative in society. But the uncontrolled excesses of society and their resulting injuries to man and nature were documented by human experience and the empirical evidence of ruined environments. The implication that should have been obvious was that the sciences of human behavior and of human ecology should advance simultaneously with the advance of physical science, medicine, and technology. To have assumed the prerogatives of a god in manipulating external nature while retaining the behavior patterns of acquisitive man could only have induced the disastrous consequences that have in fact ensued. Nature has not provided social man with homeostatic mechanisms adequate to protect him from the exponential magnification of his ecological errors through his technoscientific capabilities. He must discover, invent, or develop mechanisms for rational control of himself in utilizing the powers that he now commands. The alternative to this imperative appears to be self-destruction, for there is no evidence that "nature" can be relied upon to save society from willful tendencies long removed from "natural" control.

Not all men have been indifferent to the need for a reliable science of human behavior. Yet the need has been more apparent than the means to its realization. The reasons for lagging development and disunity among the biobehavioral and social sciences are complex. They include the psychological inhibitions that confront human efforts at self-analysis. A major element among many is the historical configuration of knowledge which divided the biological and social sciences. The locus of that ultimate knowledge which is necessary to the attainment of dynamic stability in society is very probably in the unmapped areas between biobehavioral and social sciences as we have known them.

A promising attempt to restructure the configuration of knowledge toward a more adequate science of organizational behavior has been the movement for the study of general systems. James G. Miller who has written the most comprehensive treatment of living systems to date has defined general systems theory as a set of related definitions, assumptions, and propositions which deal with reality as an integrated hierarchy of organizations of matter and energy.[42] Systems science involves a synthesis of knowledge, a multidisciplinary approach to organization phenomena, and a unified focused approach to a major aspect of nature—its systemic character.

Closely associated with systems theory has been the analysis of levels of systems development or of systems integration.[43] Through analysis of the properties of systems at as many levels of integration as can be identified, a means may be found to avoid the errors of analogy that invalidated earlier attempts to relate social and biological phenomena. The concept of a hierarchy or levels of development is not new or even recent. Contemporary emphasis has followed from the accumulation of evidence that adds empirical data to a concept that has been previously largely philosophic.

Is Social Health a Meaningful Concept?

Are homeostatic mechanisms necessary for the maintenance of social health, or is this question inappropriate because health as applied to individual organisms (e.g., humans) cannot meaningfully be applied to organizations or societies? Health is not easily defined even in individuals. It appears to be a relative concept, not often realized in an ideal or perfect state. More often it is inferred from the apparent absence of disease, disorder, or disfunction in the human body, but the criteria or tests for identifying impaired health in the body are not directly applicable to the body politique.

Nevertheless the concept of health is commonly applied to societies. Conditions identified as crime, war, economic depressions, inflation, poverty, prostitution, and homosexuality have been regarded, not without dispute, as social pathologies. But behaviors regarded as sociopathic in some societies are accepted as normal in others. Moreover, if homeostasis is a critical condition for social health, extrabiological mechanisms would seem to be required; biological homeostatic responses are confined to individuals. Whether such mechanisms or responses are, or could become, general at societal levels is as yet as conjectural as the validity of the concept of the superorganic.

There do appear to be self-maintaining mechanisms in some societies, most clearly identified in relatively small communities. Anthropologists have described their operation among various primitive peoples.[44] Whether or to what extent there may be a biological basis for a tribe's homeostatic behavior, there have sometimes been unequivocal biological consequences associated with specific behaviors which have resulted in the survival or disintegration of the community. Homeostatic responses to external and internal threats to maintenance of the community seem largely to be learned responses rather than biologically induced, but we cannot be sure that some innate biological tendency may not underlie cultural behavior. Yet a great variety of social arrangements and behaviors among people indicates that culture rather than biology is the dominant factor in the long-term maintenance of social organization.

Whether social health is a meaningful concept is a poor question for definitive answers. If sociobiologists succeed in establishing the presence of genetic tendencies and responses inherent in all humans and can further demonstrate the presence of these factors in collective human behavior then account will have to be taken of the role of biology in behavior. For the present there is little that can be said for certain, regarding social as distinguished from personal health. This is especially true for the large, complex, heterogenous societies characterizing the modern world. Within the time frame of human experience their existence has been too brief for judgment concerning their long-term viability.

The greater part of mankind's existence has been lived in relatively small homogenous groups in which homeostatic responses, cultural and possibly biology-based, can be identified. Humans are capable of living under radically different conditions, but there may be costs incurred by life in megacities and highly pluralistic societies that are cumulative and unavoidable and in the end humans may prove unable to meet. Is there a condition of societal hypertrophy (excessive growth) that in the course of time will be discovered to be a true social pathology? If societal hypertrophy is found to be inimical to human survival, can it be reversed and at what further cost?[45]

It would be too easy to dismiss this line of speculation as nonsense. Modern society has undertaken to "engineer" its future but appears to be grossly deficient in firm understanding of human limitations and capabilities. Ideological and emotional conditioning distort efforts to understand the causes of social disorder or even to achieve agreement on what constitutes disorder and whether it is socially harmful.

Health and homeostasis are social concepts fundamental in any assessment of the future prospects of modern society. They are not so much subjects for research as goals for elucidation by intermediate and more tangible objects of inquiry. If our cultural tendencies today may ultimately be blocked by biological parameters that humans cannot change, would it not be better to know this before the event rather than to learn it through experience, with all that this experience may imply? But if societal advancement were blocked by a belief that its potential had been reached, and this belief was untrue, might that not be a less desirable fate?

Notes

1. Edward Isaacson, *The Malthusian Limit: A Theory of a Possible Static Condition for the Human Race* (London: Methuen, 1912).

2. J. E. Lovelock, *Gaia: A New Look at Life on Earth* (New York: Oxford University Press, 1979).

3. *Homeostasis and Feedback Mechanisms,* Eighteenth Symposium of the Society for Experimental Biology (New York: Academic Press, 1964). For other treatments of homeostasis see the following: C. S. Koshtoyants, *The Development of Homeostasis, with Special Reference to Factors of the Environment,* Proceedings of a symposium held at Liblice, Czechoslovakia, September 1960, ed. E. F. Adolph (New York: Academic Press, 1960); A. Locker, ed., *Biogenesis, Evolution, Homeostasis,* a symposium by correspondence (New York: Springer-Verlag, 1973); L. L. Langley, ed., *Homeostasis: Origins of the Concept* (Stroudsburg, Pennsylvania: Dowden, Hutchinson, and Ross, 1973); Jules Henry, "Homeostasis, Society, and Evolution: A Critique," *Scientific Monthly* 81 (December 1955): 300–309; Dickinson W. Richards, "Homeostasis Versus Hyperexis or St. George and the Dragon," *Scientific Monthly* 77 (December 1953): 289–94; "Homeostasis: Its Dislocations and Perturbations," *Perspectives in Biology and Medicine* 3 (Winter 1960): 238–51; Anatol Rapoport, "Stochastic, Mechanical, and Teleological Views of Homeostasis," *Homeostatic Mechanisms,* Report of symposium held 12–14 June 1957, Brookhaven National Laboratory (Upton, New York: Brookhaven National Laboratory, 1958), 243–53.

4. Rene Dubos, *Mirage of Health: Utopias, Progress, and Biological Change* (Harper, New York: 1959), 41.

5. Claude Bernard, *An Introduction to the Study of Experimental Medicine,* translated by H. C. Greene, introduction by Lawrence J. Henderson (New York: Dover, 1957). See also Roger Heim, et al., *Les concepts de Claude Bernard sur le milieu intérieur,* colloque international organisé pour la célébration du centenaire

de la publication de l'Introduction à l'étude de la médecine expérimentale de Claude Bernard, 1965 (Paris: Masson et Cie., 1967).

6. Lawrence J. Henderson, *The Fitness of the Environment: An Inquiry into the Biological Significance of the Properties of Matter* (New York: Macmillan, 1913). For an updating of Henderson's findings see R. A. Bailey, et al., *Chemistry of the Environment* (New York: Academic Press, 1979).

7. Walter B. Cannon, *The Wisdom of the Body* (New York: Norton, 1932, revised 1939). The title was not original with Cannon; it was taken from Professor E. H. Sparling's 1923 Harveian Oration before the Royal College of Physicians.

8. See note 2 above.

9. T. F. Malone and J. G. Roederer, *Global Change*, Proceedings of a symposium sponsored by the International Council of Scientific Unions (ICSU) during its twentieth general assembly in Ottawa, Canada on 24 September 1984 (Cambridge: Cambridge University Press, 1985).

10. Walter B. Cannon, "Biocracy,"*Technology Review* 35 (March 1933): 203–6, 227.

11. Walter B. Cannon, "The Body Physiologic and the Body Politic," *Science* 93 (1941): 1–10. Abridged in *Scientific Monthly* 79 (July 1954): 20–26; and *The Body as a Guide to Politics* (London: Watts, 1942).

12. Jules Henry, "Homeostasis, Society, and Evolution: A Critique," *Scientific Monthly* 81 (December 1955): 300–309.

13. Jean Fernel, *De Abditis Rerum Causis* (On Hidden Causes) (Paris: C. Wechelum, 1548).

14. Sir Charles Scott Sherrington, *Man on His Nature* (Cambridge: Cambridge University Press, 1940; New York: New American Library, 1966), 6 ff.

15. Thomas S. Kuhn, *The Structure of Scientific Revolutions* (Chicago: University of Chicago Press, 1962).

16. Paul-Henri Thiry d'Holbach, *Éthocratie ou le Gouvernemente Fondé sur la Morale* (Amsterdam: Marc-Michel Rey, 1776). On the physiocrats generally see Henry Higgs, *The Physiocrats* (New York: Macmillan, 1896); Pierre Naville, *Paul-Henri Thiry d'Holbach et la Philosophie Scientifique de 18e Sude* (Paris: Gallinard, 1943); and N. J. Ware, "The Physiocrats," *American Economic Review* 21 (1931): 607–19.

17. Thomas Robert Malthus, *Population: The First Essay*, reprinted with a foreword by K. E. Boulding (Ann Arbor: University of Michigan Press, 1959).

18. A. Sergent, *"Les sciences de la vie," Combat*, 2 May 1950, 4.

19. Garrett Hardin, "The Tragedy of the Commons," *Science* 162 (1968): 1243–48.

20. Bernard Berelson, "Beyond Family Planning," *Science* 163 (1969): 533–43.

21. Herbert Spencer, *Synthetic Philosophy*. For a list of the comprising works, see "Herbert Spencer," in *Encyclopaedia Britannica*. For a critique that I regard as objective see John Dewey, "Herbert Spencer," in *Characters and Events: Popular Essays*, in *Social and Political Philosophy*, vol. 1, ed. Joseph Ratner (New York: Henry Holt, 1929), 45–62.

22. Herbert Spencer, *Principles of Biology II* (New York: Appleton, 1913), 537.

23. Ibid., 538.

24. Cannon, *The Wisdom of the Body,* 312–13.

25. Karl W. Deutsch, "Mechanism, Organism, and Society: Some Models in Natural and Social Science," *Philosophy of Science* 18 (1951): 230–52.

26. See note 23 above.

27. See note 9 above.

28. U.S. Department of Health, Education, and Welfare, *Toward a Social Report* (Washington, D.C.: U.S. Government Printing Office, 1969).

29. House Subcommittee on Science, Research, and Development, *Managing the Environment (Report to the Committee on Science and Astronautics)* (Washington, D.C.: U.S. Government Printing Office, 1968). See also Bertram M. Gross, "The Coming General Systems Models of Social Systems," *Human Relations* 20 (November 1967): 357–74. For a general treatment of indicators (there is an extensive literature) see Kevin J. Gilmartin, et al., *Social Indicators: An Annotated Bibliography of Current Literature* (New York: Garland Publishing, for American Institutes for Research, 1979), supplementing Leslie D. Wilcox, et al., *Social Indicators and Societal Monitoring: An Annotated Bibliography* (Amersterdam, New York: Elsevier Scientific Pubishing Co., 1972); and Charles P. Wolf, *Quality of Life, Concept and Measurement: A Preliminary Bibliography* (Monticello, Illinois: Vance Bibliographies, 1979). See also issues of the journals *Social Indicators,* 1977–present and *Social Indicators Research,* 1974–present.

30. Cannon, "Biocracy," 206.

31. Ibid., 222.

32. Cannon, "The Body Physiologic."

33. Koshtoyants, *The Development of Homeostasis.*

34. Cannon, "Biocracy," 227.

35. Cannon, "The Body Physiologic," 26.

36. Sergent, "*Les sciences de la vie,*" 4.

37. A. E. Emerson, "Dynamic Homeostasis: A Unifying Principle in Organic, Social, and Ethical Evolution," *Scientific Monthly* 78 (February 1954): 67–85.

38. Patrick Romanell, "Does Biology Afford a Sufficient Basis for Ethics?" *Scientific Monthly* 81 (September 1955): 138–46.

39. Ibid., 143.

40. K. E. Boulding, "Political Implications of General Systems Research," in *General Systems, Yearbook of the Society for General Systems Research II* (Ann Arbor: Society for General Systems Research, 1957), 5.

41. Sherrington, *Man on His Nature,* 6.

42. James Grier Miller, *Living Systems* (New York: McGraw Hill, 1978). Miller provides several definitions of general systems in this massive volume. For other sources see "Living Systems: Basic Concepts," *Behavioral Science* 10 (July 1964): 193–237, continued in subsequent issues. For additional references see Ludwig von Bertalanffy, "Problems of General System Theory; Conclusion," *Human Biology* 23 (December 1951): 301–12, 336–45; K. E. Boulding, "The Malthusian Model as a General System," *Social and Economic Studies* 4 (September 1955): 195–205; and J. H. Milsum, "Technosphere, Biosphere, and Sociosphere: An Approach to Their Systems Modeling and Optimization," *IEEE Spectrum* 5 (June 1968): 76–82.

43. See R. W. Gerard, "Higher Levels of Integration," in *Levels of Integration in Biological and Social Systems* ed. R. Redfield (Lancaster, Pennsylvania: Cattell, 1942), 67–87; and Alex B. Novikoff, "The Concept of Integrative Levels and Biology," *Science* 101 (1945): 209–15.

44. See Emilio Moran, ed., *Adaptability: An Introduction to Ecological Anthropology* (Scituate, Massachusetts: Duxbury Press, 1979); Richard Lee, *The !Kung San: Men, Women and Work in a Foraging Society* (Cambridge: Cambridge University Press, 1979); Betty J. Meggers, *Amazonia: Men and Culture in a Counterfeit Paradise* (Chicago: Aldine, 1971).

45. Lynton K. Caldwell, "Environmental Policy in a Hypertrophic Society," *Natural Resources Journal* 11 (July 1971): 417–426.

4
Biological Relevance of Historical Experience

There are large areas of human experience to which bioscience as we know it today cannot easily or adequately be applied or applied directly. Some fields of biology-related inquiry, for example anthropology, demography, human ecology, history of medicine and disease, and social aspects of psychology, assist understanding of human behavior, individual and social. Where applied to human experience, however, these sciences are rarely able to employ experimental methods and must have recourse to research more like that of the social sciences and humanities. For discovering biologically influenced trends in human societies, comparative studies of human experience in historical depth offer a limited but instructive resource.

Conclusions drawn from experience are inferential, and two characteristics of inferential or deductive reasoning are pertinent here. The first is that inference can be, and often has been, based on theories which distort the actual meaning of human experience and for which no practical tests are feasible. This theorizing may reflect a deficient understanding of human society and behavior, but it may also be a rationalization of evidence to support normative values. Actual experience might cast doubt upon preferred "truth." A large part of deductive political theory is, in fact, based upon revelation, inspired by moral conviction rather than by thorough and critical examination of evidence. A second characteristic, however, is that the deductive analysis of experience may sometimes be tested and evaluated by employing both conventional weight of evidence methods and, where feasible, utilizing the analytic techniques of behavioral science. This approach is more often advocated than followed for research in historical depth because there is often insufficient time and money for adequate investigation and useful results seldom can be guaranteed. To the extent that a valid reconstruction of human experience is possible it nevertheless follows that "the value of history cannot be proved—it is simply a basic fact of human experience."[1]

The Limits of Experience

The lessons of experience are limited, first by the unique nature of any event; second by the quality of information about the event; third by the capacity of persons to receive the experience whatever its form—that is, to actually comprehend it; and fourth by the ability of the receivers of information about the experience to be guided by it.

Experience and history have inspired numerous but often dogmatic proverbs and apothegms, for example, the observation attributed to Oscar Wilde that experience " . . . is merely the name men give to their mistakes." This assessment seems to be at least tacitly shared by those analysts who discount experience in the larger sense and whose faith is placed in statistical methods and in the belief that, given sufficient quantifiable data, mathematical procedures will provide a reliable basis for policy decisions. In practice, the reasonable proposition that mathematical techniques *may* help to provide basis for judgment, slips over into the less reasonable assumption that quantification alone *will* provide the base.

If the data for social choice is limited to mathematically testable evidence, large areas of experience fall outside the decision process. Not only will policy analysts fail to develop and refine methods for more adequate appraisal of unquantifiable experience, but decisionmakers may believe that they have taken all significant evidence into account when they rationalize their choices upon the limited basis of so-called hard evidence, (i.e., evidence weighed or counted).

Among his "Maxims for Revolutionists," George Bernard Shaw observed that: "Men are wise in proportion, not to their experience, but to their capacity for experience." People appear to differ widely in this capacity, thus posing a practical limit upon the uses of experience. Ability to learn from experience has a demonstrable evolutionary and survival value. The processes of learning through experience have been a major focus of biobehavioral research, and have heightened an old issue of political theory—that of the methods and ethics of humans controlling the behavior of other humans. Reinforcement through direct individual experience has been the basic idea in behavioral psychology.

But there is a discontinuity between individual motivation and collective behavior. To what extent can individual *personal* experience provide guidance for *social* policy? Obviously, the personal experience of each individual is sharply limited by time, place, and circumstances. No experience can be precisely repeated, although there are presumed to be areas of commonality among experiences described as similar. To analyze experience or to act upon it requires that capacity of which Shaw wrote, and which may be identified by two qualities of personality.

The first is openness of mind to perceivable phenomena. This implies personal qualities of alertness and awareness. These qualities further imply for full effectiveness a second characteristic, which is a range of knowledge sufficient to cope with circumstances of the environment. In any aspect of life it is easy to overlook something for which one is not looking and does not even have in mind. Breadth of knowledge becomes increasingly important when one must learn from the experience of others and especially when one moves from the personal level of experience to the social level. That knowledge is built upon knowledge is a truism demonstrated by the history of science and learning. One learns algebra before calculus. But truism or not, the usefulness of historical experience appears to be widely discounted although perhaps sometimes inadvertently among the makers of public policies. With the exception of public health, the biological history of humans has not been perceived as a reliable source of information that might guide present-day decisions or, at least, alert to possible consequences of present actions.

Granting that public policy decisions are often made under pressure and on the basis of incomplete and imperfect information, would it not be logical for policymakers to use the best information available in reaching their decisions? Such insight as we have obtained into political decisionmaking suggests on the contrary that decisions are often made with disregard for available cautionary evidence. Not only have decisionmakers been conditioned by their education and cultural milieu to utilize and select portions of available pertinent information, but they are also often prevented by institutional and ideological structuring of their roles to consider other possibilities. Personal circumstances may also affect judgments. But experience derived from direct and personal interaction with the world is necessarily microexperience. It is contemporaneous and subjective, its effect upon an individual may be biased by other experience. Hazards brought on by social behavior can be prevented only by social or macroexperience, and unless or until our capacities for forecasting are improved the source of this macroinformation is essentially historical and impersonal.

Experience as history has limitations that are readily apparent. It is vicarious and the record is incomplete and never wholly objective. At its most complete, the details and complexities of history make difficult the distillation of data that is relevant and significant to present-day issues. In the past, as in the present, preoccupation with current events obscures deeper and more slowly moving trends. And yet this liability is a reason for attention to the history of long-term social and political developments. For here, if anywhere, may be found the macroexperience of collective humanity. Here may be observed, however imperfectly, transgenerational demographic trends, movements of thought, changes in values and behavioral norms, and institutional response to these developments. For purposes of identifying the gross processes of the rise and decline, and the survival and disintegration of societies, a relatively coarse macro-overview may be sufficient; that is, if the actuality of the processes themselves has been verified. In the past, significant data have been inaccessible because the methods required for retrieval had not been developed. The sciences, and especially archaeology and anthropology, have added greatly to our ability to reconstruct the past and have thereby enlarged the sources of human experience. Written history is not therefore a finished book, but is subject, like the sciences, to revisions and to new chapters.

Discovering Relevance

History might be regarded as more relevant to the advancement of knowledge if it were regarded as source and method and not merely as historical writing. If no historical interpretations of history (i.e., histories written in the past) existed, the records and remains of the past would be the only sources for the recovery of human experience. The success and reliability of reconstructions of the past would depend greatly upon the methods and tools of recovery. Within the latest half-century, advances in scientific methods have greatly enlarged our ability to discover the character of past conditions, to identify trends and discontinuities, and to deduce cause-effect relationships. Through a variety of analytic methods in the physical and biological sciences we are able to learn more today about life in prehistoric and ancient times and chronological succession than could ever have been known before.

There are many barriers to obtaining a full and reliable knowledge of the past and yet we are able to find clues to the apparent causes of past events that may direct attention to similar or contrasting conditions in our own times. For example, dietary habits of people have been correlated with the incidence of particular conditions of health and disease. We know enough about the foods eaten by peoples in the Stone Age and antiquity to draw some contrasts with modern life. We know something about the diseases which early societies experienced and can observe changing patterns of disease over the centuries.[2] We also are able to discover changes in population densities, urbanization, commerce, agriculture, natural resources base, and in climate, and have some intimations of the ecological causes of human disasters.[3]

Demographic change may often be inferred from the physical record. Inferences can be drawn regarding relative overpopulation at particular times and places.[4] The effects of sociocultural behaviors may be more difficult to ascertain from unwritten archaeological evidence, but when added to written records and accounts, inferences may be drawn that could direct attention to possible analogies in the present. The point is not that history repeats, but that cause-effect relationships may. The historical record does indicate widespread and recurring improvidence among people in relation to their life-support bases in land and water. The phenomenon of an increasing population pressing on a diminishing resource base has occurred with enough frequency to have been more instructive than it has been.

Advances in the life sciences have provided clues to previously unexplained historical events and have suggested their relevance to modern predicaments. Population ecology is a field of scientific inquiry that offers clues to possible outcomes of demographic trends. Studies of crowding in animal populations have been cited as of possible relevance to behavioral problems of humans under crowded and confining circumstances for which we have historical information.[5] Biological significance of events ranges from the conjectural to the near certain. For example, was the decline of upperclass Roman society hastened by cooking and eating with utensils made of lead, a toxic metal that the poor could not afford (conjectural), or was the breakdown of serfdom and feudalism in Western Europe hastened by the decimation of rural populations by the Black Death (very probable)? In human beings biology and culture are fused, and only with great care, caution, and advanced methodologies have we been able to separate biological factors from other influences.

In learning from historical experience, however, it may not be necessary to know where biology ends and culture begins where both appear to be significant. Of course, to the extent that we can segregate influences we may learn better how to deal with them, including better understanding of possible synergistic effects. Meanwhile, we would be misled to mistake the boundaries of our academic disciplines for boundaries recognized in reality by nature. Human behavior and societal evolution result from interactions of biology, culture, and enviromnment and policies adequate to cope with undesired social trends must try to take account of each of them and their interactions.

Cultural changes induced or influenced by environmental change have been studied by anthropologists.[6] Whether or how biological factors have been factors in human response to environmental change is less clear than that biological and behavioral consequences have followed such changes. A cult of addictive drugs may be culturally induced, but there may be clearly measurable evidence of its biological effects. Biopolitics or biocracy may therefore be as concerned with the biological effects of culture as with the cultural effects of biology, cybernetic or feedback factors may be significant in either case.

Because we cannot investigate the living past directly we must rely heavily on written accounts and circumstantial evidence. Ecological correlations are often strongly suggestive; as, for example, in relationships between deforestation, soil erosion and exhaustion, siltation of rivers and estuaries, spread of malaria, and decline of commerce.[7] Demographic changes in societies are always of biological significance, but are usually only perceived as cultural. For example, historical experience is strongly suggestive that the integrity and stability of a society may be threatened by large infusions of foreign peoples, presence of a permanent alienated underclass, or gross generational or sexual imbalance. Political decisions might be taken to cope with the cultural consequences of these changes, but the biological consequences are not readily altered.

Experience plus science could help people anticipate trends and take appropriate action while opportunity was present. In view of well-known preferences in many societies for male children, what effect may be expected from discovery of a generally available method of gender selection at conception? What historical ecological evidence provides is indication of developments that have had confirmable consequences in the past. Today we have information and some understanding regarding trends that peoples in the past were unable to detect or to assess. We are often able to know better what happened to them than they themselves knew.

Lessons from History?

It is one of philosophy's great ironies that Georg Wilhelm Friedrich Hegel, who became the involuntary godfather of two opposing political movements each claiming to represent the fulfillment of history, wrote that " . . . what experience and history teach is this—that peoples and governments have never learned anything from history, or acted on principles deduced from it."[8] The fascist and communist ideologists who took inspiration from Hegel's philosophy of history have gone far in demonstrating the truth of his pessimistic conclusion. Yet the demonstrations are not wholly conclusive. Marxism is alive, although aging, yet continues to expand its influence. Fascism, in fact, as distinguished from political rhetoric, survives only in a comatose state. But political ideas, like the mythological phoenix, have shown the capacity to arise renewed from the ashes of disintegration. A final verdict on Hegel's political disciples must be postponed. Meanwhile, it would appear that liberal democracies also confirm Hegel's belief that peoples and governments will not learn from historical experience.[9]

Although it has been demonstrated that animals, including some people, learn from individual experience, the instructive value of history for social learning is less evident and more often questioned. Hegel did not argue that there were no lessons to learn from history; rather, that people were poor learners. He believed that history would be instructive if men approached it through the right philosophical method. He recommended a method of dialectical reasoning which Marx subsequently applied to an environment-influenced, evolutionary, materialistic theory of history. Marxist theory has been structured to rationalize the progressive realizations of Marxist values. But it cannot be said that Marxists have ignored history or have failed to find lessons in it. That Marxist theoreticians may be entrapped by their own selective perspective on history is another matter. To the extent that they broadened their intellectual horizon and critically reviewed orthodox interpretation in the light of new and unfiltered evidence, they might enhance the survival prospects for their preferred political system. But they might then cease to be Marxists.

Regardless of whether the Marxists have learned the right lessons, it is easier to demonstrate that their political adversaries have learned less, and that often of dubious validity. The liberal democracies of the West clearly did learn something from their mistakes following World War I, and following World War II they created allies out of former adversaries. But it is questionable whether their learning has been proportionate to their need. Old errors were avoided but new ones were incurred, often in apparent overreaction to perceived "sins of the past." These errors in large measure may be attributed to an essentially unhistorical perspective among large numbers of "intellectuals" in Western democracies and to the unrealistic assumptions regarding human behavior that have guided their policies.[10] These "errors" and their implications for national and social survival have been examined by Revel in his book, *How Democracies Perish* (1983).[11] Meanwhile, we will consider more closely why historical information is so often ignored or misunderstood and why so little biopolitical wisdom has been drawn from it.

Such lessons as may be learned from history are seldom made explicit by events or by those who record them. History does not speak for itself, someone interprets the evidence that is the substance and source of history. It is from this record, combined with other data and influenced by cultural predispositions that interpretations of history are made. Because human history includes all recollected human experience, its interpretation in the broad sense draws upon all fields of human knowledge, however organized. Its subject matter is the macroexperience of humanity. History is, of course, studied in particular aspects and degrees of detail. But even the biographies of individual persons and the chronologies of small communities over a short span of time must be placed in context to be fully understood. This history of microexperience may be instructive at the personal level. Persons in public life have found value in biographical works of earlier public figures. But it is erroneous to overly rely on microhistory for lessons useful to society. Lessons there may be, but their significance will almost always require consideration of the macrocontext.

Macrohistory may be the most promising way of pursuing the study of social survival. Hazards to social survival seldom emerge suddenly and without antecedents; they may be long in the making before they reach a critical point at which threatening change becomes dramatically evident. The history of social dynamics must therefore take account of transgenerational trends, movements, and processes, and to locate the interval of concern as best one can on the trajectory of history. This should not be read as suggesting that from our contemporary perspective we can obtain a cosmic view of humanity's past, present, and future. But an effort to approximate the view might greatly help to foresee and to avoid the consequences toward which events may be moving.

Linkage between micro- and macrohistory may be found through the search for indicators, for clues in the details of human experience and their environmental effects that point the direction of change. Interest in social and environmental indicators arose during the decade of the 1960s and now provides at least concepts and methodologies that could be linked through time-series to a larger assessment of historical trends. Another form of social analysis that may be given an historical dimension belongs to the general field of systems studies.

The social dynamics studies pioneered by Jay W. Forrester at the Massachusetts Institute of Technology, and the studies on limits to growth that followed from them, provide systematic and comprehensive examples of trend analysis that have influenced the attitudes of many people in public life.[12] The *Global 2000* report prepared during the administration of President Carter, drew upon dynamic systems modeling to forecast possible futures and provided a comparison with other systems models. Like historical interpretation, these projective techniques have their limitations and risks. They should alert policymakers to trends that should be examined; they are not forecasts to be accepted without question.

Although a politics of survival ought to consider a wider range of data and experience than have been incorporated in the limits studies, the undertaking of these studies and the constructive part of the criticism directed to them have provided experiments in the identification of indicators and the interactive effects of social and ecological processes that will be instructive if mankind decides to take seriously the risks that it has raised against its prospects for survival. Evolutionary biology, broadened to include its interactions with culture ought to provide clues to what social policies would enhance human well-being and survival.

Unfortunately, we are still far from adequately understanding the reciprocal influences of biology and culture. In time this must surely be a major focus of research; at present it attracts little support in the world of science. The principles of systems dynamics, broadly conceived, represents a conceptual approach to social change and evolution that could be of help in discovering the relevance of the past. The risk is that interest in method will override concern for evidence; that systems dynamics alone will be relied upon to provide a complete and coherent view of social dynamics, and that what cannot be fed into computers—which may well be most of human experience—will be regarded as inconsequential, as nondata. This risk is real in view of the tendency in modern technoscientific culture to discount the uncountable, to disregard what cannot be measured directly in the contemporary environment. It is a paradox that Western society, which of all world civilizations has been the most teleological, the most overtly guided by theories of historical progress (Christian and Marxist) should, at a critical and perhaps climacteric phase in its development, be fundamentally unhistorical in its outlook.[13] Modern rational man looks to science rather than to history for intimations of the future. But science is underdeveloped in precisely the areas in which it should be strong if it alone were to provide reliable guidance toward a sustainable future.

We may interrogate the muse of history by whatever methods we choose, but her answers will return no more than the questions and the method of questioning can elicit. Clio is not the muse of prophecy. To learn from history one must first understand what may be learned, as Clio does not overtly teach. To learn from history requires effort. History is not an open book and, in any event, reading the record is but a step toward learning—necessary but not sufficient. To learn from history is to reflect upon its possible meanings, to analyze and compare its processes, events, and outcomes. This is a qualitative task beyond the present cerebral simulations of computers which, however, may process the grist for reflection.

What strategies may prove rewarding in the interrogation of Clio? We must first identify what we need to learn. A precept of biocracy is the need to understand the factors that make for the survival of peoples, their societies, and their cultures. To understand these factors one must also understand the countervailing forces that make for decline and disintegration. Knowing what we want to learn we must then examine the record of human experience, inadequate as it may be. Here we look for indicators of trends toward societal success or failure. Equally important is identification of predisposing factors. These may be numerous, varied, and often obscure. They include cultural traits, ecological relationships, and psychophysical characteristics. And because these factors are interwoven in the texture of a whole society, one must examine its history holistically seeking out its critical aspects. The survival experience of societies may be examined as case studies, and we will now consider how they might help us formulate questions to be asked of history in search of insight into contemporary problems.

The Parable of Easter Island

It may seem contradictory to begin an examination of the uses of historical experience by reference to a culture that left no history—at least in written form. The prehistorical experience of the people of Easter Island nevertheless offers several important advantages to our purpose. It was an isolated island, a world apart from other peoples and cultures for long intervals of time; it was an island Earth, in effect a socioecological "closed system." After its settlement and until its discovery by Europeans, it was apparently free, with one probable exception, from any but nominal in and out migration. Its space and resources were finite and their limits discernible. What is known and believed regarding the history of Easter Island has been pieced together from remaining evidence. The story of Easter Island must therefore be one of inference and conjecture. But plausible alternative explanations of the evidence on Easter Island are severely limited, and there is sufficient probability of the truth of our present belief regarding its past to present as a prototype case of socioecological disintegration, of cultural history driven by biology.

Easter Island is believed to have been first settled about A.D. 460 by Polynesians from the Marquesas Islands.[14] A second wave of migrants may have arrived about A.D. 1100, possibly subduing the original inhabitants and imposing upon them a new social and religious system. Characteristic features of this new culture were the tall statuesque images quarried from stone in the interior of the island and erected, facing inland, along the coast. Reports of these statues reached Europe through Jacob Roggeveen, a Dutch sea captain who first sighted the island on Easter Day 1722. His ship's log records "remarkably tall stone figures that caused us to be filled with wonder."[15]

Fifty-two years later when British captain James Cook visited the island, all was changed. The statues were in ruinous condition. Cook declared that, "the present inhabitants have most certainly had no hand in them as they do not even repair the foundations of those that are going to decay." Many statues had toppled over and Cook found no purpose that they served in the then prevailing culture. It now appears, however, that the carving of the tall stone images ceased abruptly about 1680. What ensued may be described as a time of troubles which ended in the extermination of the old culture and its replacement by a new ideology. The chronology of events is not clear, but their sequence appears probable. We may therefore reconstruct this probability in the form of the following parable.

Polynesian colonizing explorers arrived at Easter Island around A.D. 1100. They found the island occupied by earlier settlers who symbolized their religious beliefs in crudely carved images, some of which survive. At this time the island probably possessed a substantial forest cover which provided material for the construction of houses and fishing boats. The invaders subdued the residents, reducing them to an inferior, perhaps a semislave, status. They then developed the cults of the large statues, possibly forcing the aborigines to work at quarrying, carving, and transporting the statues, which as the years passed became even taller and heavier. The largest statue attempted was never completely quarried. Measuring eighty-two feet in height and weighing an estimated four to five hundred tons it lies where it was quarried, evidently summarily abandoned when catastrophe abruptly ended the era of statue making.

The most plausible explanation of the downfall of the statue-making culture was social war, concomitant with ecological bankruptcy. During the roughly five centuries of the statue cult the population of the island continued to grow. Isolated from invasion by man or plague, the people multiplied until A.D. 1600 when there may have been 20,000 dwellings on Easter Island. If this estimate is valid the human population must have been several times this figure—an astonishing contrast to the 111 islanders remaining in 1887 after decimation by slave raids from Peru and by smallpox brought back by the few individuals who survived slavery in the Peruvian guano islands. The population today approximates twelve hundred residents.

The debacle of circa 1600 may have been a Malthusian biopolitical scenario. As the population grew, pressure upon arable land and upon sources of wood increased. Deforestation and crowding resulted, with competition for survival growing more acute. The loss of timber eventually reduced the ability of the islanders to construct fishing boats and to supplement their diet from the sea. A deficiency of nutrition and especially of protein followed. Internecine raiding for food, and resort to cannibalism, broke down whatever communalism the island as a whole may have possessed. At some point, possibly indicated by the apparent sudden ending of the quarrying and carving of statues, an uprising against the dominant order seems to have occurred. In the social war that followed, the upper class followers of the statue cult were largely destroyed and a new religion, the bird-man cult, was initiated.

The statue cult appears to have been associated with the veneration of ancestors. But they were not the ancestors of the now victorious, previously supressed classes who may have been descendants of the original aborigines. The new cult emphasized physical strength, alertness, and perhaps guile. Each year a bird-man was inaugurated—the first man to bring an egg of a sooty tern from a rocky offshore islet back to an assembly of judges on the cliffs of Easter Island.

The significance of this cult is poorly understood; it conceivably may have related to fertility. But it was unable to preserve the integrity of an Easter Island culture. For whatever reason, the population of Easter Island appears to have progressively declined, reaching its lowest point in 1887. By this time depopulation, acculturation, and poverty had destroyed all but vestigial elements of a distinctive Easter Island culture. Missionaries, traders, and Chilean officials contributed to the final process of disintegration.

The actual course of events in detail and the precise causes of the breaks in the continuity of Easter Island culture remain unconfirmed. Yet the plausibility of the parable of Easter Island is reinforced by evidence from other parts of the earth in which population and resources, notably food supply, became imbalanced to so severe an extent that ensuing social disintegration was followed by a collapse of the culture. Events ecologically similar to those occurring on Easter Island appear to have occurred among the builders of Teotihuacan on the Plateau of Mexico; and urbanized societies in Central America, the Middle East, and Central and Southeast Asia may have experienced comparable progressions of overextension, decline, and fall.[16] Again, it must be emphasized that causes and circumstances of the disappearance of the high cultures of the Toltecs, Mayas, Sogdianans, and Khmers are, at best, conjectural. What is verified, however, is that flourishing societies and cultures collapsed and their disappearance seems to have had some relationship to the overstressing or deterioration of their life-support systems. Climatic change and foreign invasion may have had a part in the disintegration, but military defeat has often been preceded by social disorder and loss of morale within the vulnerable society. Climatic change is cause-effect related to ecological change and is obviously a factor that could affect a people's chances for survival.

Reliance upon prehistoric, undocumented evidence of failure to survive is not adequate to support a theory of social causation. In a future-oriented technological society in which history is widely regarded as largely irrelevant to policy, the contemporary failures observed and recorded are far more convincing than are experiences of the remote past. The ecological behavior of society during the relatively recent past is more persuasive in demonstrating the relevance of biology to the experience of humans in history.

Examining the Record

The biopolitical thesis developed in this book was stated with eloquence and vigor as early as 1948 by William Vogt, a conservationist with extensive field experience in Latin America. His book, *The Road to Survival,*[17] was well received, even acclaimed. But its warnings of worldwide socioecological disaster were more often discounted than accepted. For this there may have been two apparently contradictory reasons.

One is that the message of the consequences of multiplying people and wasting resources became familiar to many literate and well-informed people. The message lost urgency through familiarity. The United Nations Conference on the Human Environment (Stockholm 1972), the Biosphere Conference (Paris 1968), the World Conferences on food (Rome 1974) and population (Bucharest 1974 and Mexico City 1984), and numerous unofficial symposia and seminars, gave wide exposure to the trends described by Vogt. But familiarity often breeds complacency. Conference holding is widely regarded as indicative that something is being done, that progress is being made, that resolutions and treaties signify that a problem has been taken in hand. Thus the sense of alarm and urgency that Vogt sought to impart has been lost by a generation that regards itself as sufficiently aware of the trends against which he warned.

The other reason for disregarding the prescriptions of *The Road to Survival* has been a pervasive belief among many economists, technicians, scientists, public officials, and industrialists that although Vogt's factual analysis may have had some validity, the significance that he attributed to it was all wrong. Mankind is already on the road to survival.[18] The road is called "Science and Technology." Its guide posts read: "economic development," "industrialization," "biotechnology," "agrarian reform," and "social justice." The rise of environmental awareness, and especially of a growing concern for the predicament of mankind, exemplified by the "limits to growth" thesis, challenged the expansive concept of salvation through economic growth engineered through technoscience. Spokesmen of the technoscientific persuasion undertook rebuttal of the "limits" thesis, and a new epithet, "doomsayer," was applied to environmental "alarmists."[19] In consequence the "survival" thesis appears to have become controversial rather than a subject for serious public consideration.

In principle, the survival through science thesis is plausible to persons with the mind-set characteristic of the technological society. But there are factors relating to survival of which science takes no account. Vogt's book was based on scientific evidence and was described by a reviewer as "science in the pure cause of salvation."[20] The critical difference between Vogt and many of the informed and science-oriented developmentalists lies in their assumptions. Certainly they differ in their evaluation of the evidence, but more importantly in how the predicament of mankind is to be addressed. Vogt argued for a revolutionary change of perspective. "Drastic measures are inescapable," he declared. "Above everything else, we must reorganize our thinking."[21] But although the number of persons who agree with this sense of urgency have greatly increased since 1984, the policymakers of our times see no such drastic measures as necessary or feasible.

Vogt's book was a landmark in pointing out the biological relevance of historical experience, and in this respect had a different focus, if similar thesis, from most other examinations of man's behavior in relation to his biology and environment. The survival focus impelled Vogt to move beyond analysis to prescriptive measures. He was specific about the institutional, legal, and economic reforms that would be required to put humanity on the road to survival. But, in common with almost every other prophet of eco-catastrophe and the need for behavioral reform, he was unwilling or unable to suggest *how* the measures that he saw as necessary could be brought about by political means.

The differences between the early warnings of William Vogt, Aldo Leopold, Paul Sears, and the conclusions of some of their followers in the environmental movement in the 1960s and 1970s were both methodological and philosophical. Unlike some latter-day antiscience, antitechnology, back to nature extremists, Vogt and his contemporaries based their arguments upon the findings of science regarding man's impact upon nature. The facts of physical and biological science were fundamental to their interpretation of history, and the history of human experience was an integral part of their argument. Yet their skepticism that modern society would always make wise use of science, and that science could be counted upon to retrieve errors made in its name, offered a cautionary perspective on the future in contrast to the exuberance of technocratic optimists.

Science per se was not at issue—the basic conflict was over contrasting perceptions of reality and over the need for a new view of man-earth relationships and the biological legacy of human generations. The Vogt-Leopold-Sears perspective and its successive variants saw fundamental changes in human attitudes and values as essential to the effective use of science and technology in the course of human and ecological survival. In the contrasting, and in some respects reactive, perspectives of Wilfred Beckerman, John Maddox, Julian Simon, and Herman Kahn, history, in the main, is progress toward a better world, and there seems no good reason for a basic reorientation of beliefs, values, and behaviors.

Abbreviated and simplified, the difference comes down to this: one perspective sees man as harboring within himself his own worst enemy, against which he requires moral, conceptual, and institutional protection; the other sees man as inherently innovative, resourceful, and resilient, his greatest risk being, not self-destructiveness, but rather failure to realize his true potential because of faintheartedness and nostalgia. Neither perspective can be proved right or wrong. A case can be made for each depending upon selection of the evidence. But the two perspectives cannot equally be right, and public policies may differ dramatically depending upon which perspective or combination of perspectives is accepted.

Given the inevitable uncertainties of the future, does it matter greatly which perspective is adopted, since the validity of neither can be assured? It does matter to persons to whom survival, in some sense, is a compelling value, and for an obvious reason that point will be reiterated throughout this volume. Not all policies and decisions select equally for survival. In a complex and changing world in which the relative selectivity of policies cannot reliably be foretold, a politics of survival will inevitably be cautious, avoiding the deceptive sense of security that may come from a scientific faith in statistical projections and in rigorous quantifications of data which may be of dubious relevance and sufficiency. Mathematical methods of analysis are not, per se, at issue; it is their inappropriate or exclusive application to policy questions that mislead. Uncritical reliance on tradition or common sense may equally mislead. Yet, at our present state of knowledge, analytic skill, and forecasting, it would be highly improvident to deprecate such historical experience as humanity possesses with respect to the guidance and management of its social and bioecological relationships.

By the tests of Western science the historical record is unscientific. Because it can neither be replicated nor quantified in most particulars, it may be disparaged as mere literary narrative. The "lessons" of history also allegedly fail to take adequate account of growth of knowledge on present and future decisions. It may also be assumed, optimistically, that humanity now "knows better" than to repeat the mistakes of the past. But what if human decisions and behaviors are only partially influenced by knowledge? The evidence from behavioral science depicts man as much more than a logical thinking machine and unbiased information processor.

The historical record, inadequate as it may be, provides a check upon unbridled enthusiasm for the scientific management of humanity's future. Among the high priesthood of science and technology the history of human behavior is widely regarded with condescension. Public officials and legislators, among whom there is often more misconception than comprehension of the past, tend to prefer the optimistic outlook of scientism, as long as science doesn't cost "too much" or land them in a quagmire of "controversial" issues. History, in the sense of recorded experience, may offer a most inconvenient and subversive prospect. Small wonder that men resist learning from history. Human greatness and grandeur are recorded, but history may also be a destroyer of myths and illusions, and it repeatedly documents the arrogance, shortsightedness, and folly of all races and classes of humanity and especially political shortsightedness and bureaucratic ineptitude.

The Judgment of History

Whatever beliefs, goals, or values shape man's behavior, that behavior ultimately must pass under the judgment of history. The road to extinction has been well-worn by countless species and by peoples, nations, and civilizations. Some few have taken this road voluntarily, more have been forced along it unwillingly; but the greater number have surely followed it inadvertently, not foreseeing where it would lead. The road to extinction has so often been traversed that its signs and markers should be readible, if not always clearly. But the repeated experience of humanity is that the messages are not read; they are misinterpreted, filtered, distorted, or obscured by mind-sets that transform external realities into internalized preferences. The biological and social aspects of life are so completely divorced in the minds of people generally that many might wonder why a chapter on historical experience was included in a book about the politics of biology. This book identifies indicators that could tell humanity something about its prospects for survival. One of its purposes has been to help clarify why the messages have seldom been received or acted upon in appropriate ways, and how with advances in interdisciplinary science, education, and communications they might now be better understood.

In his philosophical study on the meaning of history, *The Fate of Man in the Modern World,* Nicholas Berdyaev declares that:

> . . . what we are witnessing today is more than the end of an historical epoch—it is a judgment upon history. The results of the long processes of history have so far advanced in our day that a judgment upon the failures of history is not merely possible but inevitable. These results have been evident in national, political, and economic life. They are evident in spiritual culture as well, in literature and philosophy, for even here the

true image of man has been disturbed, the integrity of human existence is threatened.[22]

Berdyaev sees the ending of modernity as a moral judgment upon history as made by man which, in effect, is a judgment upon man, its maker. In this sense the judgment may be defined as the fate that befalls humanity as a consequence of its response to the challenge of events and of its environment. History in this context is this interactive process itself and not merely a record of the process. The standard against which judgment is rendered is essentially one of survival and success. To define this standard in terms consistent with what we may call "higher human values" is a formidable task and one that has engaged many of the most thoughtful human minds since Aristotle. This book does not contend that we are now able to adequately set such standards, but it does argue that we cannot do so without recourse to the life sciences. The question at issue is the extent to which a society realizes its possibilities and achieves such collective purposes as it articulates without destroying the basis for its continuing existence.

Because human cultures are complex and evolutionary, the judgment of history often appears qualified and tentative. It may also be absolute, often seen as "unjust," and is always remorseless. Such concepts as forgiving and forgetting are irrelevant to history which shows the implacable quality of nature because it is nature. As with the riddles of the Sphinx, only "right" answers receive a favorable judgment.[23]

History seems unjust because its judgment does not necessarily fall upon the perpetrators of social or ecological error. The rewards and penalties of history are transgenerational, each generation enjoying the benefits and bearing the costs bequeathed it by its predecessors. Efforts to compensate for the social or ecological errors of previous generations may also perpetrate injustices which however may also be understood as the judgment of history which does not distinguish personal "innocence" from "guilt" and falls upon the collectivity of a society which must bear the legacy of its past.

It is a sophistry characteristic of the modern way of thinking that because man makes history he can belatedly control its judgment or avoid its penalties. Biocratic thinking would not lead to such conclusions. But would bioecological thinking lead to a better outcome for humanity? How would the world today be different if modern man's thought and action had been guided by understanding of and respect for bioecological processes?

Had the past five centuries been characterized by ecological wisdom it is probable that at least three quarters of the people living today would never have been born. Pressure on all resources would have been correspondingly reduced because not only would the per capita demand have been less, but technology would not have been deployed in ecologically destructive ways. The human occupation of the earth would have proceeded slower and more conservatively and many plants, animals, and human cultures would have been spared the extinction that followed their contact with modernity. The incidence of social disorganizations, poverty, and many forms of disease would almost certainly have been greatly reduced as would also those forms of affluence based upon the reckless exploitation of natural resources. In all, change would have been slower, more conservative, less dramatic, and with fewer unforeseen consequences.

Obviously, one cannot be sure what a biocratic ecologically rational society would be like; no advanced civilization has produced one. As compared to modernity there would be costs as well as benefits. The fact of fewer people implies less aggregate human achievement and some loss of talent. The loss of the "good" in human populations would be offset by reduction of the "bad." But obviously no balance sheet can be drawn for so hypothetical a proposition.

It is possible, however, to defend the proposition that the lot of humanity—individually and in the aggregate—would be better than it is today in a world guided by bioecological principles. And this proposition will be more defensible if the modern era ends in the colossal tragedy that seems increasingly probable. Too many informed and thoughtful observers have foreseen the severity of the impending judgment of history on modern civilization to regard this probability as unfounded.[24] What effect this perception, presently confined to relatively few people, may ultimately have on social attitude and public policy is uncertain. Yet as historian William A. McNeill has observed, "What we believe about the past, after all, does much to define how we behave in the present and what we do towards making up the future."[25]

The judgment of history upon history as made by men is then the outcome of the choices that, in effect, a people makes collectively. Freedom to choose has been freedom to make the choices that lead to the destruction of freedom in the name of freedom. The price of freedom in a world in which all things are not possible is to make the choices that enlarge rather than reduce the scope of freedom in the long run. This means, in effect, that choices made are as important to the existence of freedom as is the actual freedom to choose. Turning again to the words of Nicholas Berdyaev:

> The age of skeptical liberty is closing, a new age is beginning. Freedom, understood as something positive and joined with creativeness, becomes creative energy. Freedom means not only freedom of choice, but choice

itself. Freedom cannot be simply a formal self-defense; it must lead to creative activity. The transition is inevitable from formal liberty, by which each protects and defends himself, to true freedom by means of which human society is creatively transfigured.[26]

This creative transformation must become the distinguishing task of policy leadership if human society is to avoid sinking into an era of disorder and deprivation. The evidence of deep troubles in the world is too strong to dismiss. History is a source of lessons and warnings as well as of judgment. Society cannot escape the judgment of the past, but it may possibly determine the judgment to be rendered in the future. The life sciences, wisely used in policy should help to ensure that history made in the future will render a favorable judgment.

Notes

1. Joseph R. Strayer, *The Interpretation of History* (Princeton, New Jersey: Princeton University Press, 1953), 14. Note also John Herman Randall, Jr., *Nature and Historical Experience* (New York: Columbia University Press, 1958). For a view of history and human behavior that defends cultural over biological influences, see Kenneth Brock, *Human Nature and History: A Response to Sociobiology* (New York: Columbia University Press, 1980).

2. See William H. McNeill, *Plagues and People* (Garden City, New York: Anchor Press, 1978), and *The Human Condition: An Ecological and Historical View*, The Brand-Lee Lectures at Clark University, 1979 (Princeton, New Jersey: University Press, 1980). Also Alfred W. Crosby, *The Biological Expansion of Europe: 900–1900* (Cambridge: Cambridge University Press, 1986); Howard C. Hopps, "The Ecology of Disease," *Pfizer Spectrum International* 19 (Fall 1976): 43–96; and Hopps, "Geographic Pathology," in *Pathology*, ed. W. A. D. Anderson and John M. Kissane (St. Louis, Missouri: C. V. Mosby, 1977), 692–736.

3. E.g., see J. Ingersoll, *Historical Examples of Ecological Disaster*, part I, HI-242-RR/A 2-3, and part II, HI-303-RR/A 1-2, (Harmon-on-Hudson, New York: Hudson Institute [now located in Indianapolis, Indiana 46206], 1963, 1964).

4. E.g., see William Vogt, *Road to Survival* (New York: William Sloan, 1940), note reading list 298–301.

5. Notably, studies by John B. Calhoun. See *The Ecology and Sociology of the Norway Rat* (Bethesda, Maryland: Department of Health, Education, and Welfare, 1963), and "Population Density and Social Pathology," *Scientific American* 206 (February 1962): 139–48.

6. E.g., Emilio Moran, ed., *Adaptability: An Introduction to Ecological Anthropology* (Scituate, Massachusetts: Duxbury Press, 1979); Moran, ed., *The Ecosystem Concept in Anthropology* (Boulder, Colorado: Westview Press, 1984); and Colin Turnbull, *The Mountain People* (New York: Simon and Schuster, 1972). See also Betty J. Meggers, "Environmental Limitation on the Development of Culture," *American Anthropologist* 56 (October 1954): 801–24; James Neel, "Lessons from

a Primitive People," *Science* 170 (20 November 1970): 815–22; and E. A. Wrigley, *Population and History* (New York: McGraw-Hill, 1969).

7. E.g., George Perkins Marsh, *Man and Nature or, Physical Geography as Modified by Human Action*, ed. David Lowenthal (Cambridge: Harvard University Press, 1965; original publication 1864); and William L. Thomas, Jr., et al., eds., *Man's Role in Changing the Face of the Earth* (Chicago: University of Chicago Press, 1956). See also William C. Bice, ed., *The Environmental History of the Near and Middle East Since the Last Ice Age* (Orlando, Florida: Academic Press, 1978); Rhoads Murphy, "The Decline of North Africa Since the Roman Occupation: Climatic or Human?" *Annals of the Association of America Geographers* 41 (June 1951): 116–32, and "Nature and Man in China," *Modern Asian Studies* 1, no. 4 (1967): 313–33; also W. S. Rosecrans, "Man and Land in History," *American Journal of Economics and Sociology* 17 (July 1958): 337–40.

8. Georg Wilhelm Friedrich Hegel, *The Philosophy of History*, trans. J. Sibree (New York: Colonial Press, 1899), 6.

9. Barbara Tuchman has described some notable examples of persistent unwillingness to learn in *The March of Folly* (New York: Random House, 1985).

10. E.g., Julien Benda, *The Betrayal of the Intellectuals* (La Trahison des Clercs), trans. Richard Aldington (New York: William Morrow, 1928).

11. Jean-Francois Revel, *How Democracies Perish*, trans. William Byron (Garden City, New York: Doubleday, 1984).

12. For a listing of Forrester's works on system dynamics, see Brian P. Bloomfield, *Modelling the World* (Oxford: Basil Blackwell, 1986), 112.

13. For a philosophical analysis of the breakup of a teleological and historical perspective on life and values in Western society, see Alasdair MacIntyre, *After Virtue: A Study in Moral Theory* (Notre Dame, Indiana: University of Notre Dame Press, 1981).

14. Grant McCall, *Rapanui: Tradition and Survival on Easter Island* (London: Allen & Unwin, 1981); and Alfred Metraux, *Easter Island: A Stone-Age Civilization of the Pacific* (New York: Oxford University Press, 1957).

15. See John Dos Passos, "Roggeveen's Misfortune," in *Easter Island: Island of Enigmas* (New York: Doubleday, 1971), 16.

16. There may be many causes of decline or extinction of societies and species. No possibly significant factors should be ignored. For an evolutionary perspective see Paul Ehrlich and Anne Ehrlich, *Extinction: The Causes and Consequences of the Disappearance of Species* (New York: Random House, 1981).

17. William Vogt, *The Road to Survival* (New York: William Sloan Associates, 1948).

18. E.g., Julian L. Simon, *The Ultimate Resource* (Princeton, New Jersey: Princeton University Press, 1981).

19. E.g., Wilfred Beckerman, *In Defense of Economic Growth* (London: Jonathan Cape, 1974); John R. Maddox, *The Doomsday Syndrome* (New York: McGraw-Hill, 1972); and H. S. D. Cole, et al., *Thinking about the Future: A Critique to the Limits to Growth* (London: Chatto and Windus, 1973).

20. Louis J. Halle, Jr., advance reader of Vogt's manuscript, *The Road to Survival*, cited on jacket of first printing.

21. Vogt, *The Road to Survival,* 285.

22. Nicholas Berdyaev, *The Fate of Man in the Modern World,* trans. Donald A. Lowrie (London: Student Christian Movement Press, 1935), 21.

23. C.f. Robert Heilbronner, *An Inquiry into the Human Prospect* (New York: Norton, 1980); and Ralph Wendell Burhoe, "The Human Prospect and the Lord of History," *Zygon* 10 (December 1975): 299–375.

24. E.g., Robert Heilbronner, *The Future as History* (New York: Harper & Row, 1968), and *An Inquiry into the Human Prospect* (New York: Norton, 1980). From a different approach see Richard K. Curtis, *Evolution or Extinction: The Choice Before Us: A Systems Approach to the Study of the Future* (New York: Pergamon, 1982); and John R. Platt, *Perception and Change: Projections for Survival* (Ann Arbor: University of Michigan Press, 1970). The greater part of this literature, however, appears in science fiction or semifiction, e.g., John Brunner, *The Sheep Look Up* (1972), and *Stand on Zanzibar* (1968); Paul Ehrlich, "Eco-catastrophe," *Ramparts* 8 (September 1969): 24–28; and George Stewart, *Earth Abides* (1949).

25. NcNeill, "What we believe about the past, after all, does much to define how we behave in the present and what we do towards making up the future" (*The Human Condition,* 75).

26. Berdyaev, *The Fate of Man,* 46.

5
Life Sciences as Problems for Politics and Law

In this chapter we are not concerned with the current state of the law regarding the life sciences advances, but rather with the conceptual and evaluative problems that bioscience presents to lawmakers. No pretense is made to an exhaustive treatment of a subject upon which books could, and have, been written. The purpose here is to help persons who are neither lawyers nor legislators to appreciate the way in which bioscience and biotechnology change the assumptions on which modern political and legal systems have been established. Ethical questions pertinent to law and justice are raised and will be considered further in the following chapter.

Throughout most of history, law was alleged to be God-given or divinely inspired and among many peoples expressed a communal consensus sanctified by tradition. Into these long-standing orthodoxies the advent of scientific rationalism brought potential discord. Faith in revealed truth was challenged by a skeptical view of all claims unsupported by tangible, testable evidence. By the twentieth century, science had already become a powerful force in society and an influence upon politics and law. Science developed its own orthodoxies regarding the tests of truth, but old orthodoxies of political ideology and religion continued to coexist.

Up to the present, modern society has generally been able to accommodate science, religion, and idealist political philosophies. Clashes have occurred, but in the main each aspect of the civic culture has operated largely within its own sphere. Creationist attacks upon the teaching of evolutionary theories have been political and legal brush fires without general threat to the established domain of science. But

---

Based in part on "Biocracy and Democracy: Science, Ethics, and the Law," *Politics and the Life Sciences* 3 (February 1985): 137–49.

this apparent accommodation between science and the civic culture may have been a truce rather than a concordat. Except for Darwinian theory, the scientific advances that attracted public attention in the nineteenth and early twentieth centuries were largely in the physical sciences. Important advances were occuring in physiology and medicine but posed no immediate challenge to social convention. The sciences of human behavior were only beginning to take form in the closing years of the nineteenth century. The attacks upon traditional orthodoxy by the rationalist philosophers of the eighteenth and nineteenth centuries were more often based on inductive reasoning rather than upon experimental science.

During the latter third of the twentieth century, science has entered a new phase which changes its relevance for politics, law, and religion. The long-awaited science of mankind is showing signs of emergence. The sciences of life have been advancing at rates without precedent. Unsettling findings and theories have been forthcoming from many of them that lay open to question some of the fundamental assumptions of civic life, of political theories, and of law. Established political and religious orthodoxies are not often explicitly or overtly challenged by propositions derived from the life sciences. Some of the outspoken defenders of nonscientific religious and Marxist orthodoxies may be found among biologists. Yet a gap appears to be widening between the belief systems prevailing in modern democratic liberal societies and the implications of the growing body of evidence regarding the nature and behavior of man, his societies, and institutions.

Hazards of New Knowledge

Religious fundamentalism or dogmatic literalism, based upon final and unquestionable revelation, cannot easily accommodate new findings regarding human behavior. The issue at debate is summarized in the question, are there some forms of knowledge so potentially dangerous that they should not be discovered or researched? If the answer is "yes," should government try to suppress or control such knowledge?

To the extent that scientific methods are accepted as the only reliable ways to discover truth, the advancement of science poses a threat to all final conclusions based upon inductive and nonverifiable propositions. The beliefs of those who are skeptical of revealed ideologies and whose understanding of science is greater than that of most people have potentially a significant impact on the role of science in social policy. These tend (with inevitable exceptions) to be the better-educated members of society, more active in civic affairs. They include scientists, physicians, engineers, and managers. These people are in general more accepting of scientific innovation than the society at large. Yet among this sector of society, there are well-informed persons who are apprehensive regarding the discovery and use of "dangerous knowledge."

Reservations regarding certain advancements in the life sciences are held by persons who see the risks and dangers as outweighing the benefits of certain innovations. There are scientists, including some distinguished biologists, among those who counsel caution with respect to dangerous knowledge.

Response to these questions normally reflects assumptions regarding the risks and ethics latent in particular developments. From a viewpoint of reason and objectivity one would expect that both risks and benefits could be estimated and evaluated. Unfortunately, in making this calculus one is dealing neither with complete knowledge nor with reliable predictions of human behavior. A consideration important to this issue is the risk of developing policies in ignorance of their possible chain of consequences for better as well as for worse. How can we tell when ignorance is more dangerous than dangerous knowledge?

There is now a large and, to some informed persons, an impressive literature regarding the predicament of modern humans, especially in relation to the effects of their activities upon the biosphere. From this perspective, the risks inherent in the present situation are so great that advancement in the biobehavioral sciences is imperative if reasonable confidence is to be placed in the survival of modern society. Insofar as this assumption prevails with respect to the biology, there is no serious talk of stopping its general progress; debate centers on particular developments which some people regard as illegitimate.

When are safeguards needed, and how should they be applied so as to be effective without obstructing the advancement of knowledge? A conclusion implicit in a large number of studies of science policy is that neither science nor society is presently well equipped to make this determination. The exclusive and, to the public, often arcane character of science has been much criticized but nonetheless persists. Some scientific propositions are intrinsically difficult to comprehend, but failure to understand science as a mode of learning is largely a failure of education. Some colleges and universities, with no more than modest assistance from the National Endowment for the Humanities and the National Science Foundation, and almost none from private philanthropy have attempted to broaden the base of public understanding of science and its social relevance. In engineering, law, medicine, business, and public administration there have been professors and practitioners who see their fields of learning as inadequately prepared to deal with the interrelationships of professional practice and social policy as they are affected by the advancements in the sciences. Their viewpoints are in the minority, however, and have not been shared by discipline-bound colleagues.

The state of science in the abstract has rarely been a principal issue anywhere in politics. But the low level of public attention to science, per se, is no measure of its impact upon public affairs. Public officials can seldom wholly escape responsibility for interpreting socially significant scientific information. In scientifically advanced countries, institutions of public policymaking and justice must now deal with an increasing number of issues involving the life sciences and public policy, and they often encounter difficult questions for judgment. Lawyers, judges, and administrators may not be the least-informed members of the public regarding science and its implications, but the prevailing logic of the law and politics and the logic of science are based upon different premises. Moreover, in cases in which scientific information is important, public officials have been handicapped generally by the conspicuous lack of attention that science was given in their educational preparation for public careers.

One answer to this deficiency has been the proposed "science court." The idea of a special court or courts to consider cases in which science is a critical factor received a flurry of attention in the law reviews in the mid-1970s.[1] But the idea has not caught on, perhaps, among other reasons, because of difficulty in identifying the cases in which science would be a critical factor, and the unsuitability of the adversary process in resolving scientific questions. Moreover, since the life sciences and biotechnology reach out and touch every facet of the lives of people, it becomes increasingly difficult to mark off an area of the law that could be considered primarily a case of science policy. Although the factual content of a controversy may sometimes be clarified by science, the human concerns and values at stake reflect normative ethics. Scientific facts are seldom absolute or invariant, but ethical propositions are far less objective; they are reflective of normative, subjective perceptions.

Law and policy regarding acceptable risk have been thrown into confusion by the inability of society to reach a consensus on criteria to reconcile the benefits of advances in the life sciences with the enlargement of individual or collective human rights. Escalating litigation and insurance rates provide tangible evidence that prevailing ethicolegal assumptions regarding risk do not always operate to protect the public interest. They may also lead to gross inequities in the resolution of liability claims, although their rationale is to obtain just and rational allocation of responsibilities for demonstrable injury. Nevertheless, in an era of ever more sophisticated technologies there is good reason to move away from a legal philosophy of caveat emptor. Government, and law, have accordingly been increasingly involved in protection from injury to individuals and compensation for injury if and after it occurs.

Rational Basis of Law and Equity

Advances in the life sciences present society with difficult problems that the legal system is asked to resolve. Unfortunately, the confusing and often obstructive practices that are too often tolerated in the legal system compromise the ability of modern society to accommodate to advances in the life sciences. Law and the sciences do not rely primarily on the same tests of truth. In arriving at legal conclusions, the law takes into account not only facts garnered from the "real world," but also assumptions which, by some scientific tests, could be considered irrelevant. In the United States, for example, scientific analysis of the properties of marijuana and tobacco would hardly justify the legal distinctions between them. To lawyer, judge, and police officer it is the political fact and not the pharmaceutical or botanical fact that makes the difference. But scientific information can influence political response to changing public attitudes as shown by the growth of restrictions on smoking.

There is a growing set of examples of current interpretations of the legal concepts of negligence and liability which pose a threat to advancement in many areas of medicine, pharmacology, and bioengineering. Large awards granted to litigants in cases alleging injury or death from use of a new pharmaceutical product lead people to infer that equity requires a risk-free world, although the evidence of the sciences indicates that such complete freedom cannot be attained. A more basic problem is in the application of strict liability to the manufacture and sale of pharmaceutical products that may benefit the great majority of users, but may harm an unpredictable minority; what is the most equitable and also socially preferable way to help unwilling victims without discouraging generally beneficial innovation? The future of the law in pharmacology and medicine will remain uncertain until society reaches a more settled and sound consensus on the nature and limits of acceptable risk, and on the implications of the doctrine of strict liability in the field of biotechnical innovation.[2]

Law extends to a large number of matters in human affairs beyond the scope of science. In many of the social issues which politicians, judges, and public administrators must confront, science is singularly unhelpful. Nevertheless, the welfare of human society is not well served by policies that ignore relevant evidence. Controversy, however, is often over both the facts of an issue and over the way in which those facts should be interpreted. Science can estimate the life prospects for a severely handicapped infant but cannot say if life-support technologies should be maintained.

The argument that law and policy in a democracy represent popular will, might be questioned by comparing scientifically constructed and administered public opinion polls with what the judges and administrators declare the laws to be. Yet the more fractionated the state of public opinion, the more latitude and less guidance available to judges and bureaucrats. But latitude does not ease the exercise of their responsibilities, rather it makes their roles more difficult.

The present relationships of science to the law allows lawyers and judges latitudes which some of them find convenient for policy activism but others find stressful in their search for "sound" judgments which may require a major effort to understand highly complex issues. For others, judicial opinion seeks a defensible compromise between deeply divided value positions in society, both of which find support in law and precedent. The latitude following from uncertainty allows official action to respond to political expediency, to the personal biases of legislators and judges, but also to seek sufficient common ground to allow contending factions to live together (i.e., a modus vivendi). But the life expectancy of scientific facts is in itself uncertain; the law can hardly be certain if it relies more heavily on the state of scientific opinion than its definitiveness would warrant.

How has the law responded to rapid, unforeseen, and unprecedented changes in society brought on by scientific innovation? Confusedly and unevenly, but it should not be expected to exhibit behavior radically different from that of the society of which it is a part. Not all of the social changes of the past half century are attributable directly to science or technology, although a very large number of them may be, and a larger number have been, indirectly influenced. The courts have responded both positively and negatively to the importunities of political activists for some biopolitical issues while avoiding others that might be regarded as no less important.[3]

With respect to behavior regarding sex, race, health, and the environment, the law has been invoked to change popular behaviors and assumptions. In none of these topical areas, except, to some extent, the environment, has scientific knowledge (too often too narrowly selected) provided reliable guidance for public policymaking although facts alleged to be scientific have often been invoked. In the case of the environment, science has been a moving force in many public policies, particularly in respect to those that are health related. Many environmental policies, however, reflect aesthetic preferences and moral attitudes independent of scientific knowledge. Social science has been cited in decisions relating to the effects of discrimination on social and other grounds in education and employment. But the proper limitations of the courts to the particular issues in the cases and controversies before them may lead to nonscientific distortions if translated into social policy without qualification or broader considerations.

With respect to the biocracy-democracy tension, an important question concerns the extent to which nonscientific values are consistent with scientific knowledge of the real world. Almost any matter relating to genetics seems vulnerable to political attack; the misuses of genetics have been well advertized. Yet sexual relations and human reproduction, traditionally joined together by religion, have in recent years been put asunder by lawyers, legislators, and judges acting under the influence of findings by biologists and behavioral scientists. To the conservative mind, the law in many of these issues has been excessively responsive to science. The published work of Alfred C. Kinsey may have done more to change public policies toward sex than any rethinking of sex-related issues by legalists.[4]

Whereas "liberalization" has been the general rule with sex, caution has been the tendency with respect to the more esoteric subject of genetics. The biotechnology of conception control has certainly been a factor in the liberalization of sex relationships and their decoupling from human reproduction. Interpersonal discrimination on grounds of sex or race, or who engages with whom in sexual relations, have a present salience which the judiciary feels obliged to recognize. But who reproduces and gives birth to future generations is a very different question, one which lawmakers generally prefer not to consider. In 1987, however, the Roman Catholic Congregation of the Doctrine condemned the greater number of the new reproductive technologies and called upon governments to adopt restrictive measures.[5] In March 1987, a controversy regarding surrogate motherhood (the case of Baby M) received nationwide notoriety at the same time as the Catholic pronouncement. Politics as well as females are being fertilized by the new technologies. Genetic technologies have already been before the courts and are thus well within the domain of politics.

Eugenics may return as an issue after the decline of this once popular study nearly half a century ago. There are persons, some well-informed in the life sciences, who reject genetic engineering as applied to humans and who believe that influencing the genes of future generations should be left to nature or to God. Some scientists conclude that humanity cannot improve upon natural selection and efforts to improve the human genetic endowment (i.e., eugenics) will invariably subvert science. The perceived threat of eugenics has now reappeared indirectly through sociobiology, no matter how sincerely leading sociobiologists may protest that they have no eugenic intentions. More significantly, a neo-eugenics movement has indeed apppeared. It does not necessarily follow that eugenics today will fall into the errors or perversions of the past, but there is also no guarantee that it will not lead to socially devisive issues. Further developments in the science of human genetics and a major change in social climate would be necessary before an explicit commitment to eugenics could gain social or scientific legitimacy.

Genetics, per se, may be neither good nor bad, yet as understanding of the genetic factor in human characteristics, health, and behavior increases, along with the ability to manipulate these factors, there would appear to be an inevitable public expectation that this knowledge be put to the service of human welfare. There is already apprehension that this new knowledge opens a Pandora's box of troubles. Genetic considerations (i.e., heredity) are hardly new in human affairs or in social policy. They are present by inference in the folkways of many primitive and traditional societies. Whether eugenics in its historical sense has a future is conjectural. However, research in human genetics seems likely to discover options for genetic control not heretofore present. In some form the objectives underlying the more sober concepts of eugenics seem likely to persist.

There does appear to be a strong emotional predisposition among contemporary intellectuals and ethicists to attribute human differences almost wholly to environmental influences. Acceptance of a genetic basis for inequality would not serve the ideology of the political left or of equalitarian democracy. Although the genetic basis of human personality is immensely complex and may never be fully understood, environmental influences are likewise poorly understood, and persons have often transcended environmental influences that would seem to present unsurmountable barriers to personal development. As biologists have often pointed out, environmental and genetic influences interact; all organisms have genes and all genes exist in some environment. Genetic mutations may be induced by environmental factors. No one, including the antigeneticists, has satisfactory explanations for why some persons overcome adverse environmental influences and others do not. It seems reasonable to suspect that genetics could be a factor.[6]

If the message inferred from certain prominent geneticists that genetics has little to do with ordinary human affairs, should be sufficiently modified by future findings, a restructuring of popular beliefs and attitudes might well follow and these would not necessarily be harmonious with liberal democratic assumptions. Few Americans have thought of the future in genetic terms, or of what social and political differences might follow should they do so. It is probable that so-called genetic engineering in the public interest, intended to phase out genes regarded as harmful to future generations, would affect some persons far more than others and hence is discriminatory in a socially unacceptable sense. There is certainly conflict between some of the prevailing human rights concepts in contemporary society and a vision (as distinguished from the reality) of a genetically sanitized "society" in the future. It's plausible to suspect that the passion with which some geneticists deny the social relevance of their science is driven by a subliminal fear that genetics may prove to have a powerful relevance for social policy in the future and that this policy may not square well with a highly individualized concept of human rights or with ideological versions of a "just" society. On the other hand the optimistic views of eugenecists may fail to take account of the immense complexity of genetic heritage.[7]

The tempo and scope of change have something to do with how biopolitical issues and relationships develop. There is a normal tendency in societies, and certainly in the United States, to improve and modify social policy without adopting or even considering an overall guiding rationale. Debate there often is, yet like Rabelais' comment on the Athenians, there is a tendency never to consult until after the decision has been taken. This practice may be regarded as a practical necessity in pluralist democracies in which extended consultation may widen differences to a degree that prevents governmental action.

Changes in public policies toward smoking and abuses of the environment illustrate the tendency of American social policy to advance incrementally with relatively little philosophical rationalization. This may be socially desirable, avoiding a politics of ideological confrontation. Nevertheless, some opponents of proposed changes in policy believe they see a long-range master plan or metapolicy behind these measures. The restrictions against smoking represent to them a step forward in the march of certain groups in society toward reshaping the behavior of their fellow citizens and are, as such, contrary to some interpretations of human rights and traditional American liberties. More explicitly, opponents of environmentalism detect a conspiracy of right or left to throttle free enterprise and put an end to economic growth or to indulge the aesthetic preferences of the affluent at cost to the poor. These and more "elitist" objectives, according to conspiracy theory, are being concealed from public scrutiny by the plausible front of concern for the environment.[8]

In the absence of an overall political philosophy, into which they might be integrated, many of these incremental disputes over health, behavior, or the environment have taken the form of "single issue politics." Where positions are sharply contradictory, as they are between the zero population growth and right to life movements, there is no apparent basis for political reconciliation. For the most part, politicians have not seriously attempted to reconcile differences; this could expose them to the risk of pleasing no one. They have either chosen to evade conflict, or like Ronald Reagan, to satisfy selectively and primarily "rhetorically" a constituency against legal abortion and for the unqualified "right to life" of the unborn. An incremental approach to policy on any matter (not merely biopolitical) enables candidates for public office to select and to modify their positions without too obvious an appearance of inconsistency. Difficulties may arise for politicians, however, where various incremental measures gradually are perceived as adding up to some more general overall guiding philosophy of society.

Biocracy Reshaping Law

To attempt a point by point inquiry into inconsistencies between the state of the law and life science knowledge would hardly be feasible. Neither science nor the law are sufficiently exact for precise analysis; for every apparent inconsistency one could expect to find numerous qualifications. A more productive approach is to identify those biocratic principles with which the law should be consistent insofar as it endeavors to be consistent with reality as revealed by science. Obviously, there are science-related issues in which science cannot help law. To the extent that biologic principles can be empirically verified and are widely accepted, the law, where contradictory, could be modified in a manner consistent with democratic values. Biocratic principles and propositions, although based on biology, are not, nor cannot be, wholly derived from scientific knowledge or method. Questions of ethics are also involved, although how the answers are to be discovered is often far from evident.

Science provides a generally reliable method for ascertaining the probable consequence of some forms of human action. History, as noted in the preceding chapter, is not always reliable but is nevertheless an often useful guide to foresight. Cause and effect relationships in human affairs cannot be measured with much exactitude, yet large effects may be reliable indicators of the consequences of human behavior. For all its deficiencies, however, history records a vast amount of human experience and is not without lessons that are consistent with what modern science could predict. Demographical and ecological aspects of human history were formerly given more attention than they are today, greater reliance now being placed upon science. Experimental science, still largely reductionist, has not cast much light on major behavioral trends, but paleoecology and anthropology provides some clues to human relations with the natural world in the past.[9]

The successes of the life sciences have been largely at the micro level, and appear to have influenced behavior more often than explained it. Estimates of the possible results of proposed actions can be compared with one another and with a range of human values to ascertain degrees of acceptability among available alternatives. Analytic techniques employing scientific methods have merged during the past two decades in the form of environmental impact analysis, technology assessment, and cost-benefit and risk analysis. These and other science-based techniques for prediction and evaluation have been built into the statutory law of the United States and of other countries. They have become widely accepted instruments of policy, but it would be unrealistic to expect that they would have achieved, within a relatively few years, complete acceptance or full effectiveness. Nevertheless, evidence indicates that they have had an incremental impact upon law and public policymaking although they have not thus far been aggregated to offer society a new biocratic paradigm into which legal principles could be reconfigured.

Medical and pharmaceutical products have been subject to increasing degrees of testing and control since early in this century. A broad range of regulatory measures affecting food products and environmental sanitation are the result of scientific methods of analysis and testing, and have obtained almost universal public acceptance. Very few people would now regard the enforcement of public regulations on sanitation as an infringement upon the rights of restaurant owners and food handlers. None, I believe, would regard them, in principle, as "undemocratic." Until the past decade it seemed to be a settled point of law that government, on behalf of the welfare of society, could quarantine or otherwise restrict the freedom of carriers of disease; however, as of the 1980s, the state of the law (at least in the United States) is not so certain.

It is not clear whether or not courts of law, with what some people might deem an excessive concern for private rights, might permit the unrestricted freedom of a suspected carrier of an infectious disease.[10] Chemical plants which sought to bar child-bearing women from certain occupations that might be harmful to unborn children have been told by environmental activists that factories not safe for everybody are not sufficiently safe for anybody.[11] To selectively bar women from any employment situation has been alleged to be an unacceptable infringement upon their civil rights. Paradoxically, it does not appear that civil rights activists regard the right of a prospective mother to smoke during pregnancy, now clearly established as harmful to the fetus, to be an infringement of its civil rights. To be consistent, should not the activists insist that pregnant women abstain from smoking, inasmuch as exercise of the right to smoke endangers the right to health of an unborn child?[12] Lawsuits for "wrongful life" and associated torts relating to human reproduction may presently clarify or confuse this issue.[13]

Humans are culture-bound and culture-forming animals; science and logic may be a part of their cultures, but are never all of it. Scientific findings, which address only a part of human concerns, are seldom definitive in shaping social policy regarding those concerns; intuitive judgments and value preferences are of equal or greater influence. Biocratic assumptions are derived from biology as second order inferences based upon a probabilistic inference which must be evaluated on the basis of human experience and checked against findings of other sciences, including the often less reliable sciences of human society. Biology informs us of how the living world works and of what to beware, but does not tell us how to put biocratic principles into effect. That is as much a task for the humanist as for the scientist.

Efforts to deduce principles of law and ethics from observed nature are as old as political philosophy, but such inferences do not provide an adequate theoretical foundation for the concept of biocracy. These naturalistic or organismic theories, like the previously noted political and religious dogmatic ideologies, were based upon inferences as often inductive as deductive. They were not derived from controlled observation, from experiment, or from an adequate evaluation of all relevant evidence. Attempts have been made to deduce theories of ethics and governance from biologic principles and processes; "dynamic homeostasis," described in chapters 3 and 6, was identified by biologists Walter B. Cannon and Alfred Emerson as a possible basis for a new ethicopolitical order. Present lack of popular or scientific acceptance of these ethicopolitical propositions is not necessarily a measure of their ultimate significance. Although they did not succeed in closing the gap between social theory and biological knowledge, they have added connective links that someday may be joined to form a coherent biosocial theory of human behavior. Meanwhile the impetus toward a coherent biosocial ethical theory appears to come more from the biocrats (i.e., biologists) than the democrats (i.e., politicians).

Nonetheless, the life sciences do fundamentally affect what many people believe regarding the nature of life and the ethical basis upon which human attitudes and behaviors toward life should be judged. Some of these attitudes and behaviors may conveniently be categorized as: (1) conceptualization of life, (2) evaluation of life, (3) freedom and responsibility, (4) justice and equity, and (5) safety and survival. There may be biocratic principles implicit in these topics that cannot be clarified in advance of developments of life science knowledge. Even so, inferences may be drawn from the present state of science that suggests the possibility of significant changes in the future state of ethics and the law.

Conceptualization of Life

How one understands and defines life is basic to a broad range of social attitudes and public politics. The most basic biocratic principle is that human life is an integral part of the living systems of the biosphere, that humans are unique elements in the biosphere although not unique in all respects as compared to other forms of life. Self-reproducing, self-organizing cellular material is the common denominator of all living things. Life, therefore, is living material and is not to be confused with "a life," e.g., a human individual. Moreover, a human life in embryonic form may be distinguished from a human person.

The question "when does life begin?" posed by some members of Congress with a view to writing a law is poorly phrased; the concern is not really with when the phenomenon of life began.[14] Science cannot answer that question today, although evidence points toward the emergence of life on earth more than three billion years ago. There is an ecological right to life viewpoint illustrated by Christopher Stone's book *Should Trees Have Standing?* (1974)[15] However, in the United States the political debate over the right to life that has agitated Congress really concerns only human life. The question which some Congressmen are really asking in an unsophisticated way is "when does the state of being a human person begin?" These lawmakers propose to define the term life by statutory law; for example, at the moment of fertilization of the human egg. But of course that egg and the fertilizing sperm are also alive. Search for the beginning of life leads to infinite regression. What the right to life includes, and how it is defined, has implications for a number of other troublesome policy issues, including birth control and abortion, euthanasia, capital punishment, wildlife management, and military conscription. The single track perspective of the right to life lobby appears to be blind to the implications of its objective.

How life is conceptualized is fundamental to how one views public policies regarding severely handicapped or defective unborn children. It also affects policies of care and treatment of individuals physiologically alive but mentally dead, aged persons in the terminal stages of illness or the not so aged persons with the presently incurable condition called AIDS. It also has priority implications for the allocation of funds for research basic to the cure and prevention of disease as against their allocation for noncurative treatment, e.g., kidney dialysis or the artificial heart. When Governor Richard Lamm of Colorado raised the question of justification for the artificial prolongation of life of certain aged and terminally ill patients relative to other medical priorities, his remarks set off a chorus of disapprobation with a minor counterpoint of approval.[16]

It is clear that a biologically informed concept of life is at odds with a great many prevailing social assumptions.

Evaluation of Life

The way in which the varied aspects of life are evaluated follows from the way in which life is conceptualized. But human attitudes are frequently paradoxical. Within the range and diversity of opinion, perspectives on life tend to focus on either human-centered or life-centered values, although these simplified, contrasting viewpoints are not necessarily mutually exclusive. A homocentric or human-centered view tends to regard life, essentially human life, as set apart from the rest of the living world. There is of course more than one homocentric viewpoint. Although in theory theocentric or God-centered, the orthodoxies of Judaism, Christianity, and Islam, in practice, tend toward homocentrism. More explicitly homocentric are the deistic philosophies of the Enlightenment and the atheistic ideology of Marxism/Leninism. Homocentrism puts human interests above all else in nature, although it is not necessarily solicitous about the welfare of all human life. In the biocentric perspective, humanity is not necessarily less valued but is valued differently and seen as belonging to the evolving order of the natural world.[17]

Quality of life considerations might logically have a larger place in the biocentric perspective than in attitudes less regarding of the natural world, although this might not always hold for humans if their quality of life (by some definition) were to be constrained for the benefit of other species. Biocentrism tends to evaluate human life not only in terms of what it means to be human, but also in terms of basic concepts of life and of those human characteristics that link them to, as well as distinguish them from, other life forms. The homocentric view that human life is under all circumstances more valuable than any other manifestation of nature contrasts with the biocentric emphasis on the quality of human life within the context of life-giving nature. Pure homocentrism recognizes no rights in nature, but only rights of humans against nature. Rights are a cultural concept, hence, from a biocentric viewpoint, there is no rational or authoritative basis for the proposition that humanity has a right to crowd all other living things off the face of the earth.

Conversely, do species have a right to life—a right *not* to be eliminated? A tough ethical question arises here. Do all organisms equally share a right to existence—including the cholera bacillus? Is there a "rightness" in the maintenance of the planetary biosphere and its interrelating systems, as distinguished from individual or species rights? The concept of an ethic based on dynamic homeostasis, as proposed by biologist Alfred E. Emerson and in chapters 3 and 6, suggests a way in which this question might be answered.

Justice and Equity

If humanity has a right to take what it needs or wants from the natural environment, could the failure of nature to answer to human demands be regarded as unjust? There is massive confusion in human society between alleged rights of humans to life and health, and the consequences of disregarding the "laws" under which the biosphere is "governed." Life is often judged to be unjust or unfair when perhaps it should be seen more realistically as tragic. Injustice and unfairness imply moral lapses which may not be applicable to the acts of people who, through necessity or lack of understanding, destroy the ecological base of their livelihood and suffer the consequences.

Beliefs regarding justice and equity have been influenced by assumptions regarding reciprocal obligations of humanity and nature. Were nature created for human needs and convenience, nature's failure to provide food, fiber, water, and living space might be regarded as unjust, were it not also that humans, as "owners" and managers of land and other resources, are intermediaries between nature and other humans and their actions may not always be perceived as separate from those of nature. The emperor of ancient China was believed to play such an intermediating role; if famine, flood, or other natural disaster struck he was seen to have lost the "mandate of heaven" and faced overthrow. Administrators of the Soviet Union risk loss of their mandate (i.e., their jobs) if crops fail. Presidents of the United States have been held responsible for hard times on the farms.

The modern response to recalcitrant nature is to force productivity through technology. Success in agriculture, in the harvesting of natural resources, and in the exploitation of nonrenewable sources of energy gives credence to political arguments that if nature fails to meet human needs (or demands) social institutions are to blame. Nature cannot be held accountable for disaster, but public officials may be held responsible for failure to protect people against devastation by severe storms, floods, insect infestations or the spread of disease.

Biocratic perspectives are no more consistently "fair" or "equal" at the individual level than at the societal level. Many individual capabilities are very unequally distributed. Native endowments work to the advantage of some individuals and to the disadvantage of others. Biological diversity has been the outcome of the evolutionary process and it has resulted in incommensurate characteristics not only among species, but within species. Nature is not concerned with inequalities; humans are, particularly since, from a moral perspective, inequalities are seen to result in inequities. Humans have sought ways to transcend natural inequalities through religion and politics. The equality-equity issue may in time be clarified by more adequate understanding from the life sciences, but values widely regarded as democratic (among others) may thereby be brought into question, or in the equity-equality context require further definition.[18]

As more is learned about the consequences of human exploitation of the living world and about the impact of human behavior and genetic factors on the quality of life of succeeding generations, transgenerational justice and equity may receive more consideration. Do humans have a right to transmit damaged or lethal genes to future generations, or to exploit the natural environment so as to leave only damaged and impoverished ecosystems to their descendants? Does a right exist in any meaningful sense, only if those who claim it are prepared to defend it? If natural rights are not defended by nature, in what sense are they rights? Regardless of their political unattractiveness, these questions accompany advances in the life sciences. Biocratic solutions, if sufficiently disseminated and accepted throughout society, may lead to social interference with traditional "inalienable rights"; however, new rights may arise in other contexts.

Freedom and Responsibility

The progressive liberalization of human societies since the eighteenth century has seldom received instruction from the life sciences although surely affected by their advances. "Liberty" is an appropriate concept in the life sciences in only a contextual sense. All freedom known to the sciences exists within various constraints; none is unlimited. Time is an inexorable limit on freedom. As observed in the discussion on justice and equity, individuals have limited freedom to choose what nature provides conditionally. Where choice is relatively unrestrained, there is an implied responsibility on the part of whomever has made the choice for the consequences which follow for better or worse. The meaning of "responsibility" could be modified to the extent to which a radical sociobiology might demonstrate genetically predetermined behavior. Such a development, however, seems unlikely. In any event the issue of determinism in relation to freedom and responsibility may not have much practical significance in ordinary human affairs. Yet deterministic biobehavioral theory could influence, and to some extent has influenced, attitudes toward deviant or "criminal" behavior. How one understands deterministic influences and responsibility may also have a bearing on attitudes toward poverty and toward alleged societal deviation.

Responsibility is a uniquely human concept. The issues it raises are systematically evaded in those modern societies in which social conditioning rather than innate character is widely believed to determine human conduct. Responsibility may be retained, even emphasised as a concept, but it is achieved through indoctrination or discipline in a hierarchy. The association of the concept of responsibility with class-structured, authoritarian, prescriptive societies has made it unpopular since the emergence of "liberal individualism" and of "social liberation" movements. Revolutions, wars of liberation, mass protests, revisions of the laws, and introversive life styles may change relationships among humans; they do not change the basic relationships of humans to the natural world. So-called conquests of nature are misconstructions. In actuality, the phrase describes the use of some principles of nature to offset, not repeal, the effects of others. Thus the so-called conquest of space requires rigid adherence of basic laws of physics and biology; in no fundamental way is humanity liberated from dependence upon the biosphere. It has become increasingly clear that the irresponsible uses of knowledge and techniques developed in physics and biology can be dangerous to society and to its future. Yet is it not their misuse, inspired by doctrine, rhetoric, ambition, and fear that are dangerous, and not the knowledge per se?

Experience in the sciences teaches that nature, properly understood, responsibly managed, and its conditions respected, can to some degree protect humanity from hazards inherent by natural forces. This conditional freedom does not exist apart from knowledge responsibly understood and applied. Modern theorists of political economy have had relatively little to say about this aspect of human responsibility. Inappropriate rhetoric has praised "irrepressible human will and indomitable human spirit" rather than advocating the informed and responsible behavior required of humans as citizens of the biosphere.[19]

Safety and Survival

Biocracy, as a mode of perception and evaluation, helps to clarify the human predicament in the natural world, to identify the biological alternatives available to human society, and to estimate the probable consequences that could be expected to follow from choosing among those alternatives. To speak of survival, risks the accusation of hyperbole; yet the destruction of many cultures, regimes, and societies has been preceded by unwarranted confidence in their survivability. That extinction is a common fate of species is a lesson written large in biology and hardly less large in the history of culture. It appears that humanity has no divine dispensation against such fate. Considering the overwhelmingly life-denying state of the geophysical universe, life as a phenomenon must be viewed as highly improbable and exceedingly precarious. The alleged toughness and persistence of life are dependent variables within the very special environment of planet Earth. The cockroach is far better equipped to survive unaided on the planet than is humanity.[20] The principal survival problem for humanity within the present state of the biosphere often appears to be the perversity of humanity itself.

Yet time and again, humans and their societies find themselves locked into behaviors and dependencies from which they cannot escape and which will destroy them if they do not. The dilemma of modern economies powered by the burning of coal, oil, and natural gas is a dramatic case in point. Residual gases from this combustion may drastically alter atmospheric and biological conditions on Earth, but modern society has no way to free itself of its dangerous dependency other than to find another source of energy which does not also entail environmental hazard. This, modern society has not yet been able to do.

Discipline and responsible behavior can be exacted from a relatively contained, cohesive society if the necessity is apparent. Sparta offers an example. Although culturally unattractive, it does not appear that Sparta's austerity led to its ultimate downfall. Whether mass democracies can exercise sufficient collective restraint to avoid self-destruction has yet to be seen; it is no more certain whether biogenetic factors are in any way involved, but that possibility, however apparently remote, ought not be dismissed without consideration.

And Toward What Conclusions?

It has been widely argued that neither human society nor democracy will be secure until fairness, equity, and justice are achieved everywhere throughout the human community. Biocracy raises doubt as to how far these qualities or values, as conventionally understood today, can fully prevail in any society. Their present interpretation does not derive from science, which opens to question ultimate social consequences should biobehavioral research undermine expectations that these ideal values, as presently understood, can ever be achieved. If biocracy does not negate these values then why, if their achievement by humans continues to be unreached, are humans unable to live up to their declared principles?

In an effort to more fully understand cultural values, further inquiry into their applicability to the noncultural world of nature would be useful. If human society were in all respects unique in the natural world, such comparisons would have little meaning. However, while humans are unique in degree of mental and cultural development, the life sciences are revealing more and more commonalities with other living species. The sciences of ethology and behavioral psychology, for example, have identified patterns of animal behavior which were thought to be unique to humans. Even warmaking, as Jane Goodall has demonstrated in her studies of chimpanzees, is no longer to be regarded as uniquely human.[21]

There is in the field of political thinking today a philosophical gap between traditional values and the findings and implications of advances in the life sciences, on which no unifying structure has been constructed. In the absence of a rational and coherent synthesis of these elements, there is a lack of credible alternatives to presently established political theories. Is it because the state of life science knowledge is not adequate for such synthesis? If such is the case, is it possible to identify those areas of attention of biobehavioral science where further development is especially needed? Why, at the course of the most spectacular era of advancement of knowledge, beginning with Darwin's publication of *On the Origin of Species by Means of Natural Selection* (1859), has no comparable advancement in the understanding of politics and behavior occurred? Is it a reproach to the leadership in social and political thought that near the end of the most creative century in human history, no really new concept of governance has gained general credence? A century of advances in a broad range of sciences and technologies has added little more than technical proficiency to the processes of government, for worse as well as for better.

It may be argued that if new political ideas were really needed they would have emerged. Yet both history and evolution indicate that neither species nor societies are necessarily conscious of what they need to survive; or, if aware, that they can acquire what is needed where and when they need it. The argument for biocracy rests upon the proposition that a viable future depends upon adherence to the basic principles derived from the life sciences, as mediated by human values, and tested for their real-life consequences. The argument has become critical because only now has human society acquired the scientific knowledge and the tools to permit it deliberately to influence its biological future. Whether it possesses sufficient social knowledge to take up this option with wisdom is a very different question.

What can be done to enhance the prospect for survival of the constructive achievements of humanity and of humanity itself? Dramatic changes seem necessary yet improbable; nevertheless it would be presumptious to regard the human prospect as hopeless and so the social implications of biocracy merit attention. Disagreement exists over the degree of urgency attributable to the effects of biology on politics, law, and public policy. Nonetheless, the issue is of sufficient importance to humanity's future to justify special consideration in two areas. Both may be described as "educational."

The first is increased interdisciplinary inquiry, largely through research, into biological influence on human behavior. This effort would include, but not be limited to, sociobiology, since a number of different modes of inquiry are needed to encompass the full range of biobehavioral interactions. Sociobiology, as defined by its practitioners, is one of these, but should not occupy the field of inquiry to the exclusion of other approaches. Nor should the field of inquiry be restricted because of a priori objections to the consideration of controversial questions such as the genetics of intelligence or the biological modification of behavioral tendencies. The success of interdisciplinary investigation at the interface of biology and the social sciences requires that opportunities for extended education and interaction among the relevant disciplines be increased and encouraged in universities and other research institutions.

The second area of attention involves discourse on the subject of biopolitics. Informed discourse is a condition of nearly all socially constructive achievements. Uninformed discussion is unproductive, even counterproductive. Mere opinion is not enough. Joint exploration and collective learning is a necessary preliminary to more general public involvement in the consideration of biopolitical issues. Beyond the initial exploratory stage, this intellectual interaction might direct its attention to certain laws and to patterns of social behavior, with a view to examining the consequences of existing ways of thinking and acting. This effort could be initiated by colloquia or seminars including, among others, biologists, social scientists, ethicists, and lawyers. Investment in focused and informed discourse now might prevent or diminish destructive conflict in the future.

It appears today that a major advancement in human social capability is needed if the human species is to survive misfortunes of its own making. Two thousand years of philosophy and religion have not enabled humanity to avoid international and social war, environmental contamination, resource impoverishment, overpopulation, and social degradation. Better ways of organizing humanity's perceptions of its own circumstances, its real alternatives and the probable consequences of alternative choices are needed. Biocracy may be one way.

Notes

1. See *Proceedings of the Colloquim on the Science Court*, sponsored by the Task Force of Presidential Advisory Group on Anticipated Advances in Science and Technology, Leesburg, Virginia, 20–21 September 1976 (available from National Technical Information Center, doc. no. PB 261305); Task Force of the

Presidential Advisory Group on Anticipated Advances in Science and Technology, "The Science Court Experiment: An Interim Report," *Science* 193 (30 August 1976): 652–56; Arthur Kantrowitz, "The Science Court Experiment: Criticisms and Responses," *Bulletin of the Atomic Scientists* 33 (April 1977): 44–48; Harold N. Kern, "Law, Fact, and Science in the Courts," *Columbia Law Review* 46 (June 1966): 1080–1116; Lee Lovinger, "Science and Legal Thinking," *Federal Law Journal* 25 (Spring 1965): 153–66; and Charles M. Overby, "Science, Technology, and the Judiciary: Some Problems and Possibilities," paper prepared for American Society for Engineering Education, Annual Conference, 27–31 June 1971, Grand Forks, North Dakota (bibliography appended).

2. See James F. Childress, et al., eds., *Bioethics Reporter: Ethical and Legal Issues in Medicine, Health Care Administration, and Human Experimentation* (Frederick, Maryland: University Publications of America, 1983-1986); *BioLaw: A Legal and Ethical Reporter on Medicine, Health Care, and Bioengineering* (Frederick, Maryland: University Publications of America, 1986), looseleaf; I. P. Cooper, *Biotechnology and the Law* (New York: Boardman, 1982–present), looseleaf. See also Walter Wadlington, Jon R. Waltz, and Roger B. Dworkin, *Cases and Materials on Law and Medicine* (Mineola, New York: Foundation Press, 1980); Margaret Gruter and Paul Bohannon, eds., *Law, Biology, and Culture: The Evolution of Law* (Santa Barbara, California: Ross-Erikson, 1983), see review commentaries in *Politics and the Life Sciences* 3 (August 1984): 108–19; Marc D. Hiller, *Medical Ethics and the Law: Implications for Public Policy* (Cambridge, Massachusetts: Ballinger, 1981); and William A. Thomas, ed., *Science and Law: An Essential Alliance* (Boulder, Colorado: Westview, 1983).

3. On a biopolitical complaint in which the courts found no merit, see "Court Rejects Rifkin in Biotech Cases," *Science* 235 (9 January 1987): 159.

4. Alfred C. Kinsey, Wardell B. Pomeroy, and Clyde E. Martin, *Sexual Behavior in the Human Male* (Philadelphia: W. B. Saunders, 1948); and Alfred C. Kinsey, Wardell B. Pomeroy, Clyde E. Martin, and Paul H. Gebhard, *Sexual Behavior in the Human Female* (Philadelphia: W. B. Saunders, 1953).

5. "Instruction on Respect for Human Life in Its Origin and on the Dignity of Procreation: Replies to Certain Questions of the Day." For comment see *Newsweek*, 23 March 1987, 42–43; *Time*, 23 March, 1987, 58–59; *New York Times*, 16 March 1987, late edition, 1, B5.

6. For a view of the genetic basis of inequality, see R. J. Williams, *Free and Unequal: The Biological Basis of Individual Liberty* (Austin: University of Texas Press, 1953), and *Biological Individuality: The Basis for the Genotrophic Concept* (New York: John Wiley, 1963).

7. See Theodosius Dobzhansky, *Genetic Diversity and Human Equality* (New York: Basic Books, 1973); and Richard C. Lewontin, *The Genetic Basis of Evolutionary Change* (New York: Columbia University Press, 1974). Lewontin has been one of the more trenchant critics of eugenic theory. For social and political attitudes toward genetics and eugenics see J. E. Meade and A. S. Parkes, *Biological Aspects of Social Problems*, a symposium held by the Eugenics Society in October 1964 (New York: Plenum Press, 1965), 226; Loren Graham, "Science and Values: Eugenics Movements in Germany and Russia in 1920s," *American*

Historical Review 82 (December 1977): 1133–64; and Donald K. Pickens, *Eugenics and the Progressives* (Nashville, Tennessee: Vanderbilt University Press, 1968). The issue of "Lysenkoism" during the Stalinist era in the Soviet Union is a notorious instance of the political manipulation of genetic theory and application with implications for eugenics; see David Joravsky, "The Lysenko Affair," *Scientific American* 207 (November 1963): 41–49. For some earlier American views on genetics and eugenics, see Horatio Hackett Newman, *Evolution, Genetics, and Eugenics*, 2d ed. (Chicago: University of Chicago Press, 1925). For a more recent and generally unfavorable perspective see Mark H. Haller, *Eugenics: Hereditarian Attitudes in American Thought* (New Brunswick, New Jersey: Rutgers University Press, 1963, 1984); and Daniel J. Kevles, *In the Name of Eugenics: Genetics and the Uses of Human Heredity* (New York: A. A. Knopf, 1985).

8. The principle exponent of the conspiracy allegation is William Tucker. See *Progess and Privilege: America in the Age of Environmentalism* (New York: Doubleday, 1980), and *Environmentalism: Does It Require Regulation* (New York: Praeger, 1985); also, Richard Neuhaus, *In Defense of People* (New York: Macmillan 1971); and M. J. Grayson and T. R. Shepard, *Disaster Lobby: Prophets of Ecological Doom and Other Absurdities* (Chicago: Follett, 1973).

9. E.g., Paul S. Martin and H. E. Wright, *Pleistocene Extinctions: The Search for a Cause* (New Haven: Yale University Press, 1967); Richard B. Lee and Irven De Vore, eds., *Man the Hunter* (Chicago: Aldine, 1968); Richard B. Lee, *The !Kung San: Men, Women, and Work in a Foraging Society* (Cambridge University Press, 1979); Richard B. Lee and Irven De Vore, *Kalahari Hunter Gatherers: Studies of the !Kung San and Their Neighbors* (Cambridge, Massachusetts: Harvard University Press, 1976).

10. See articles by Sandra L. Schultz, "Social Equality and Infectious Disease," and Baruch S. Blumberg, "Social Responsibility and Genetic Disease," in Robert H. Bland and Masako N. Darrough, eds., *Biological Differences and Social Equality* (West Port, Connecticut: Greenwood Press, 1983). Note also "Helen," a report on prostitution and Aquired Immune Deficiency Syndrome (AIDS) on Columbia Broadcasting System's "60 Minutes," 19 February 1984. The American Civil Liberties Union has opposed the mandatory testing of prostitutes for AIDS.

11. Oil, Chemical, and Atomic Workers v. American Cyanamid Company, 741 F. 2d 444, 339 U.S. APP.DC 222, D.C.Cir, (22 August 1984); and Wright v. Olin Corp., 585 F. Supp 1447, 767 F. 2d 915, 4th Cir. N.C., (31 August 1984).

12. See Robert H. Blank, *Redefining Human Life: Reproductive Technologies and Social Policy* (Boulder, Colorado: Westview Press, 1984), 131–51.

13. See Turpin v. Sortini, 182 Cal. Reporter 337, 643 p. 2d 954 (3 May 1982); Harbeson v. Parke Davies, Inc., 98 Wash. 2d 460, 656 p. 2d 483 (6 January 1983); and Procanik v. Cillo, 502, A. 2d. 94, 26 N.J. Super. 270 (20 August 1985).

14. See Jesse Helms, "To Provide That Human Life Shall Be Deemed To Exist from Conception," S. 158, 19 January 1981, 97th Cong., 1st sess., and later, the Jepsen-Helms bill, S. 467, 1983. For comment on the original bill, see *Congressional Quarterly*, 28 February 1981, 383–87, and 25 April 1981, 729–30.

15. Christopher D. Stone, *Should Trees Have Standing: Toward Legal Rights for Natural Objects* (Los Altos, California: W. Kaufmann, 1974); and Paul Shepard,

"Animal Rights and Human Rites," *North American Review* 259 (Winter 1974): 35–42.

16. Governor Lamm's remarks made on 29 March 1984 were widely noted in the press; e.g., *Time* 123 (9 April 1984): 68, *U.S. News and World Report* 96 (9 April 1984): 18, *American Medical News* 27 (20 April 1984): 18.

17. For the significance of a biocentric versus a homocentric viewpoint, see Paul W. Taylor, "In Defense of Biocentrism," *Environmental Ethics* 5 (Fall 1983): 237–43, and "The Ethics of Respect for Nature," *Environmental Ethics* 3 (Fall 1981): 197–218. Also Richard A. Watson, "A Critique of Anti-Anthropocentric Biocentrism," *Environmental Ethics* 5 (Fall 1983): 245–56.

18. See references to Williams, note 8 above.

19. Cf. Julian Simon, *The Ultimate Resource* (Princeton, New Jersey: Princeton University Press, 1981).

20. E. Larson, "Why the Cockroach May Thrive Another 350 Million Years," *Wall Street Journal,* 14 May 1984: 1, 17. See also M. H. Nitecki, ed., *Extinctions* (Chicago: University of Chicago Press, 1984).

21. Jane Goodall, *The Chimpanzees of Gombe* (Cambridge: Harvard University Press, 1986); and Frans de Wall, *Chimpanzee Politics: Power and Sex among Apes* (New York: Harper & Row, 1982).

6
Biocratic Interpretations of Ethical Behavior

Our concern here is with the influence of life science knowledge on concepts of ethical conduct; not with analysis of ethical conduct. Throughout the chapters of this book there are frequent references to relationships between biocracy and ethics.[1] In this chapter those relationships are made more explicit.

Before moving into into the discussion, three points should be made clear. First, every political system or ideology rests upon ethical assumptions and seeks to justify its objectives in ethical terms. Even political doctrines, which some people regard as amoral, contain an implicit ethic consistent with doctrinal premises. Second, ethics in this volume are understood as reflective and evaluated judgments upon human behavior. They are products of recollected experience and thus tend to be contextual and evolutionary rather than universal and absolute. There are acts which under any circumstance almost all reasoning people would regard as absolutely bad. But the broader principles of ethics tend to be more complex and situational, and their implications change and develop over time. Third, it follows that political conflicts often involve conflicting ethics—conflicting beliefs regarding what is right and what is wrong. Political rhetoric characteristically reveals ethical postulates that are defensible only through recourse to faith, tradition, or doctrine. This is notoriously so in biopolitical issues in which advocates of opposing policies regarding population levels, immigration, abortion, euthanasia, drug use, and genetic engineering put their ethical positions beyond debate as unequivocally right, both as to the interpretation of actual facts and to moral judgment concerning their significance. Democratic politics, insofar as it is a politics of compromise, does not easily accomodate these issues.

A basic difficulty in dealing with ethical positions on biopolitical issues is their conceptual incommensurability. For many issues there seems no rational way to attain moral agreement among conflicting claims and so no generally acceptable solution to conflict is possible. This circumstance is further developed by Alasdair McIntyre who writes:

> . . . the rival premises are such that we possess no rational way of weighing the claims of one as against another. For each premise employs some quite different normative or evaluative concept from the others, so that the claims made upon us are of quite different kinds . . . for example, premises which invoke justice and innocence are at odds with premises which invoke success and survival.[2]

It follows that some biopolitical issues cannot be resolved by rational or political means. The theory of biocracy is that attitudes and practices over time will alter the perceptions and beliefs of people so that cultural change rather than rational persuasion or law may be the ultimate way of accommodating social values to new life science knowledge.

Ethical Basis of Politics

In order to understand the role of ethics in the attempted resolution of biopolitical issues, one needs a more general and in-depth background understanding of the relationship between ethics and politics.

In the earliest formalizations of public philosophy, ethics and politics constituted unitary systems for the guidance and evaluation of human conduct. In the *Ethics and Politics* of Aristotle and the *Analects* of Confucius, the political order is, or should be, the expression of social arrangements and behaviors believed to be ethical. The legitimacy and survival of a political system or regime depended therefore upon its exemplification of the moral principles that united the society. Whatever the practice may have been, political philosophies of antiquity—in Greece, China, India, and Rome—described the ideal political order as an institutionalized system of ethics. For example, the "legitimacy" of the Emperor of China required possession of the mandate of heaven, and public disasters might signify that the mandate had been lost through royal misconduct.

Doctrinal separation of ethical and political theory appeared with the rise of transcendental religions, notably in Christianity. But the separation was not thoroughgoing. In the lives and teaching of particular individuals, ethical conduct was often sharply demarked from the political affairs of this world. But religious leaders often sought either to create a theocratic polity on earth or to persuade secular government to enforce the doctrinal, including ethical, prescriptions of their faith.

In modern Westernized societies the separation of religion in the generic sense from politics is far more apparent than real. What appears as a decline of religious faith may at least in part be seen as a transformation. Beginning with its so-called Enlightenment of the eithteenth century, the focus of religious concern among the educated classes shifted from heaven to earth, from God's will to the people's will, from righteous conduct to social justice. In contrast to medieval Christianity, modern Christian concern shifted from theology to ethics. The concept of Christendom faded with the Reformation and the gradual admission of Jews into full participation in Western society. The Judaic emphasis upon social justice appears to have reinforced this trend in Christianity so that by the twentieth century the Judeo-Christian ethic, rather than traditional Christianity, provided the religious-philosophical foundation for liberal humanitarian and socialist thought in the West. And Arnold Toynbee is surely correct in describing Marxism as a heretical version of the Judeo-Christian ethic.

In Islam, ethics and politics were always a unity. Islamic law was derived, at least formally, from the Koran, and the judges and administrators of Islamic regimes in theory simply applied the principles revealed by Mohammed. From its very beginning Islam was both worldly and otherworldly; the Islamic state was the worldly expression of the faith. Thus modernization placed severe strains upon Islamic political systems, as in departures from the Islamic code by Amanullah in Afghanistan and the Pahlavi dynasty in Iran.

The background of traditional relationships between politics and religion created a problem for the political expression of ethics that has not been satisfactorily solved in any modern society. The problem may be best understood when expressed as a question: What is the source of ethics? This question leads to others, for example: What legitimizes the application of particular ethical principles to an entire society or to all humanity? How, if at all, can one determine which among conflicting ethical principles is the better or the more valid? What determines which ethical principles are approved and promoted by governments?

The philosophers of the Enlightenment were unable to agree upon the answers to these questions. The growth of science complicated the problem. The most common approach to a solution was through a deistic view of the cosmos that identified God with nature. Human affairs were ordered by the laws of nature and of nature's God. Natural law, which natural man knew intuitively, was the authoritative source of ethical principles. The advancement of science was expected to confirm the moral judgments of the Enlightenment. But Van Rensselaer Potter has noted "how little science was available to Kant and other philosophers of the Enlightenment. They were so stimulated by the advances made by Copernicus, Galileo, Kepler, and Newton that they felt confident that the workings of the whole world would soon be revealed."[3]

Science as known in the eighteenth century was dominantly physical and mechanical. The sciences of biology and behavior were not to take shape until the end of the next century and not to mature until another century had passed. Thus the philosophical foundations of modern political ethics were laid by thinkers who repudiated the "superstitions" of the past, widely accepted an essentially mechanistic interpretation of nature, and had little valid knowledge or understanding of the biological basis of human conduct.

Generalizations concerning the conceptual basis of political ethics are what the word generalization implies, a statement frequently but not universally true. One should always recognize that there are particulars that contradict or qualify the general. There are always exceptions to generalizations and some may be more significant in the long run than the general rule. Thus the place of nature and the influence of science in biopolitical ethics have included paradoxical and even contradictory elements.

The search for a rational basis for Christian ethics led to nature. But nature as perceived in the Enlightenment was only in part derived from scientific information and that part was heavily weighted by astronomy and physics. Newtonian mechanics clearly influenced opinion regarding the constitutional structure of government and the organization of society.

This concept of the systemic unity of nature was conventional in the thought and writing of intellectuals in the eighteenth century. In the *Essay on Man* (1734), described as a book of "ethic epistles," Alexander Pope observed how "worlds on worlds compose one Universe . . ." and " . . . how System into System runs."[4] And following the concept of the great chain of being he wrote that:

> From Nature's Chain whatever link you strike, Tenth or ten thousandth, breaks the chain alike, And if each System in gradation role, Alike essential to th'amazing Whole; The least confusion but in one, not all That System only, but the whole must fall.

"Order is heav'n's first law," the poet declared, exemplifying the prevailing notion that although the world and nature were conceived holistically, the principles governing that unity were essentially hierarchical and mechanistic. The dynamics of nature were cyclical rather than evolutionary.

In effect, eighteenth century political philosophers saw rational government as embodying principles of order, balance, subdivision, and systemic unity. These qualities were believed to be inherent in nature and were extended to human society as a part of nature.[5] The Constitution of the United States of America, adopted in 1787, is the most notable practical example of rational government based upon natural principles. Historical experience was searched for guidance in the structuring of political institutions. But history was not viewed as alternative to science, but rather a source of examples of the workings of nature in human society.

Conceptual Sources of Bioethics

Tracing the historical development of any complex idea ultimately becomes an exercise in simplification as the significant is separated from the incidental. One may trace the growth over time of the idea of evolution and of its influence on ethical concepts. But the details, variations, and contradictions of real life cannot be wholly encompassed in the narrative without loss of coherence and sense of direction. Description of the emergence of bioethics in modern society may be adequate for purposes of understanding public attitudes without being complete in all respects.

Van Rensselaer Potter in his book *Bioethics* identifies his subject as "ethics based on biological knowledge."[6] The goal of this knowledge (for he regards it as purposive) is a wisdom which he describes as a "science of survival." Among the observations that could be made about bioethics we will confine our attention to four.

The first is that bioethics is an identifiable body of concepts and writings emerging as a corollary to evolutionary theory. Bioethics as a practical philosophy of survival and a catalyst of political thought begins with the controversy over Charles Darwin's *On the Origin of Species by Means of Natural Selection* (1859). The relevance of bioethics increased dramatically with the emergence of a science of genetics after 1900 and of ecology by midcentury and with the extensions of biological knowledge into practical affairs through biotechnology. By the last third of the twentieth century, bioethics had become an established interdisciplinary field of study.

The second observation is that for all its modernity, the roots of bioethical thinking extend far into the past. We have previously noted (chapters 2 and 3) the influence of biological concepts upon social theory. And just as biology throughout most of history was not a distinct science or discipline but rather an aspect of philosophy, so beliefs about nature and its processes influenced ethics per se and did not constitute a subdivision or special field of ethics. Bioethics is therefore a consequence of the growth, specialization, and practical application of knowledge. And this specialization has occurred not only in the sciences, but also in philosophy, where refinements and abstractions of thought among professional philosophers have frequently handicapped communication with the general literate public. A consequence has been that bioethics has developed without as well as within the field of formal philosophical ethics. Its proponents have largely been biologists and not philosophers. Professional ethicists have sometimes regarded the incursion of scientists into ethics much as many scientists have regarded the intrusion of philosophers and theologians into science (with dismay). But with the "explosion" of biological knowledge and biotechnology after 1950, bioethics has increasingly become a public concern, with active involvement of people lacking credentials either as scientists or philosophers.

A third observation is that the immediate lineage of bioethics is multidisciplinary. This seems obvious as the name itself suggests a hybrid. But whereas the ethical content of bioethics is often more inclusive than the ethics studied by most philosophers, the biological content is much less than the whole of biology. Antecedents of contemporary bioethics may be found specifically in the sciences of genetics (including evolution and eugenics), in physiology, in ecology, and in the behavioral sciences of psychology and ethology. We will return presently to the respective influences of these disciplines upon bioethics.

The fourth observation is that bioethics provides the rationale or distillation of biology-derived ethical principles underlying biocracy and the politics of survival. Biocracy does not rest upon the Judeo-Christian ethic, and although compatible with some of it, is not a derivative of it, and is at variance with certain of its commonly understood tenets. Although some part of the literature of bioethics attempts a reconciliation of scientific knowledge and Judeo-Christian ethics, conventional, institutionalized Christianity may be regarded as one of the two major barriers to bioethical behavior; the other being found in Toynbee's Christian heresy, Marxism. Neither of these ideologies appear to be unequivocally incompatible with bioethics; their antagonism is located in their more fundamentalist and dogmatic expressions. Bioethics would seem to have no place in fundamentalist Islam. Its presence in other religious ideologies of Asian origin is equivocal but not in principle incompatible.

Returning to the conceptual origins of bioethics, the genetic-evolutionary source may be traced to publication of *On the Origin of Species by Means of Natural Selection* in November 1859. In a historical sketch of the progress of opinion on the origin of species, Darwin himself outlined briefly some of the contributions to evolutionary thinking made previously to the publication of his work.[7] It is apparent that evolutionary thinking had for some time been "in the air" and the principal significance of Darwin's formulation lay in its effective and timely formulation of the thesis along with its subsequent use as a referent or point of departure for the development of thinking about the process of organic evolution.

Prior to publication of Darwin's work Herbert Spencer, who must be regarded as one of the forerunners of behavioral science, had applied an evolutionary hypothesis to the development of society. As Darwin personified anti-Christ to fundamentalist theologians, so Spencer became a *bete noire* to liberal humanitarians and socialists. Both men were widely misunderstood and misinterpreted. Hostility to their ideas was directed primarily to the mechanisms through which they believed that organisms and societies evolved and differentiated. In chapter 3 of the *Origin,* entitled "Struggle for Existence," Darwin wrote: "I have called this principle by which each slight variation, if useful, is preserved, by the term Natural Selection, in order to mark its relation to man's power of selection. But the expression often used by Mr. Herbert Spencer of the Survival of the Fittest is more accurate, and is sometimes equally

Spencer developed his concept of "fitness" in *Principles of Biology* (1864). His ideas varied in some respects from those of Darwin, notably in the belief that alterations in an organism acquired through adaptation to the external environment might be genetically transmitted. But it was his belief that political intervention in the evolutionary processes of nature was harmful to the human species that aroused the anger of socialists, liberals, and humanitarians—few of whom appear to have thought it worthwhile to consider Spencer's theories on survival within the broader context of his synthesis of the sciences. The fitness concept and its implications at once became, and continues to be, a basic point at issue between evolutionists and humanitarians, with science interpreters such as Thomas Henry Huxley and his successors attempting a formulation of the evolutionary hypothesis that would reconcile some of the opposing opinions.

Advancements in genetics and in cellular and molecular biology extended and complicated the evolutionary issue by introducing the possibility of mankind assuming control of its own evolutionary mechanisms or, under theories of sociobiology, being controlled by them. Evolutionary thought had undercut the thesis of special creation; genetics appeared to threaten the foundations of liberal equalitarian society. This threat was potential, not imminent; but the very thought that biological science might undermine the philosophic assumptions of liberalism and equalitarianism was sufficient to arouse bitter controversy and to cut short the early growth of a science of eugenics.[9]

Premature, exaggerated, and misused claims for eugenics (the planned genetic improvement of the human species) placed it outside the scope of establishment science from the mid-1930s onward. Emotional reaction to the racial myths of National Socialist Germany, and ideological disapproval by Marxist theoreticians placed eugenics along with parapsychology in the limbo of academic respectability. It is doubtful, however, that the pariah status of eugenics in principle is perpetual. Genetic research continues to function as a conceptual time bomb which, at some unannounced future date could conceivably, and contrary to the wishes of geneticists, blow a large hole in conventional beliefs regarding the character, limitations, and distribution of human qualities and abilities. Although human genetics has become one of the major focuses of bioethical concern its ultimate implications for bioethics is still symbolized by the question mark.

If evolution is the mother of bioethics, physiology is its father. A seminal treatise in physiology was Claude Bernard's *An Introduction to the Study of Experimental Medicine* (1865). Bernard's contribution to bioethics was indirect, but fundamental; it was a basis for the concept of dynamic homeostasis in organic life. As interpreted by Bernard, life existed in two complementary environments: "a milieu exterieur in which the organism is situated, and a milieu interieur in which the tissue elements live." The primary condition for the freedom and independence of existence was the stability of the internal environment in response to changes in the external environment. This adaptive characteristic of the milieu interieur was subsequently described by physiologist Walter B. Cannon as "the wisdom of the body," and the adaptive process identified as dynamic homeostasis.

As we shall presently see the concept of dynamic homeostasis is one of the more important contributions of physiology to bioethics. We have previously noted (chapter 3) how zoologist Alfred E. Emerson proposed it as a unifying principle in organic, social, and ethical evolution, and the foundation of a scientific ethics.[10] Maintenance of harmonious, systemic order has been an important attribute of naturalist theories of ethics. And the concept of unifying principles among various types and levels of systems was influenced by the work of Ralph W. Gerard, a physiologist and one of the founders of general systems theory. But again it should be noted that these inputs to bioethics were in the tradition of organismic interpretations of social structure and behavior, and were an extrapolation of organic order and function in physiology to the functional roles and cooperation of people in society. For example, Friedrich Engels in his *Dialectic of Nature* (1898) followed Darwin in conceptualizing "the universe in its totality as composed of ascending levels of organized matter—from molecules as it were to human mentality."[11] Engels believed "that if dialectical laws could be verified in nature, then the validity of such laws in history would, as a consequence, have to be admitted; the truth of the dialectics in nature entailed the truth of Marx's concept of history."[12]

Closely related in effect to the influence of organismic or systematic biological concepts on ethics has been the demonstrations of organic interrelatedness, complementarity, and reciprocity provided through ecology. Of the several divisions of biology, ecology is the most general and the most syncretic, extending into behavioral and social science. Although ecology is a science, its concepts also have been expressed in ethical and aesthetic terms sometimes verging upon religious mysticism: Liberty Hyde Bailey's *The Holy Earth* (1915) or Rene Dubos' *A Theology of the Earth* (1969) are examples. And the ideal of an ecological ethic runs through the writings of Aldo Leopold, Loren Eisely, Rolf Egberg, and the poetry of Robinson Jeffers. There has surely been more writing on ethical aspects of human relationships to the environment than on other bioethical relationships.

Finally, bioethics has roots in psychology going back at least to Herbert Spencer, and with nourishment variously derived from the writings of the Gestalt and Behaviorist schools and from the Depth Psychology of the psychoanalysts. The direct relevance of psychology to bioethics is most apparent in relation to theories of personality and to perceptions of and attitudes toward man's environment. Writings of Kurt Lewin and of ecological and biological psychologists have refined our understanding of the relationship between the mind of man in his milieu interieur and his existential external environment.[13]

The influence of behavioral science on environmental ethics has perhaps been as important for the nontruths that it has exposed as for the new truths that it has discovered. In this respect anthropology has also affected bioethics in displaying the broad array of ethical beliefs associated with man's understanding of nature and human behavior. Thus both psychology and anthropology have helped us to understand *how* people conceptualize their relationships to their environments, including other people and the rest of nature. They have also helped us to understand somewhat better than before, the biological basis for individuation in human personality. For bioethics imply conscious and conscientious individuals—products of biological and cultural evolution and capable of classifying attitudes and acts as right and wrong, wise and foolish, with qualifications regarded as appropriate.

The influence of instrumentation, of tools and techniques, on knowledge and belief should also be appreciated. The science and technology of optics has changed man's view of his environment philosophically as well as physically. The telescope and associated instruments of astronomical reckoning were indispensable to construction of the systemic, orderly, celestial mechanics that provided an analog for social and political philosophy in the seventeenth and eighteenth centuries. In the nineteenth century the microscope provided insights that modified earlier mechanistic theories. In the twentieth century electronic computers not only amplified the potential utility of other instruments but supplemented the human mind in analytic thought.

Principal among these was the ability of computers to store, retrieve, and combine vast quantities of information and to do so at speeds greatly exceeding human retentive and computational capabilities. Development of computers capable of simulating real life developments and of projecting interacting trends into the future, provided a means by which people might test values and alternative courses of action by the probable consequences of their realization. For some areas of policy it has become possible to foresee the probable consequences of policy before taking action. The effectiveness of computers in forecasting is, of course, limited by the quantity and quality of available information. Computers cannot solve the ethical problems of mankind, but to the extent that reason prevails in human society, they are a powerful force for reality-oriented choices.

Consistent with its foundations in knowledge rather than in inspiration, bioethics does not provide answers to all bioethical questions. And in common with some aspects of the most rigorous of the sciences and mathematics, bioethics rests upon some assumptions that are not susceptible of proof. A basic bioethical assumption is that, in principle, life is good. But life pervades the biosphere of which life of the species and each personal life is a part. All life, moreover, exists at costs that cannot be avoided. Hence assuming that survival or preservation of life is good; any particular act or event in support of life or survival may not therefore be good. Bioethics as ethics of survival are situational within a holistic and evolutionary context. They avoid the dangers of abstract morality by direct linkage with empirical fact. From the sciences of paleontology, evolution, and biochemistry we learn that life develops under conditions that appear to be very exceptional in nature; and the continuation and development of life depend upon the fitness of the physical environment for life and a paralleling adaptability of life to changes in that environment.

We have learned that organic evolution represents a long series of experiments most of which end in failure—if that is an appropriate way to describe the extinction of species and ecosystems. Taking survival of the biosphere and of the phenomenon of life to be an absolute good, and of an individual human life to be a conditional good, a bioethical view regards life-respecting and life-sustaining behavior as good in principle. But from the sciences of ecology and physiology we learn that life is characterized by competitive struggle, and that while mutual accommodation and coexistence are also characteristic, the system of nutrition upon which all life is sustained involves sacrifice of life. The concept of the food chain might be repugnant to some deductive principles of ethics, but no ethicist has survived except at cost to other life.

The concept of "right to life" in the abstract therefore has no bioethical significance. No such right exists in nature. At best, only opportunity is available, provided nature's ground rules are observed. Human life is temporary and subject to termination by causes beyond the reach of law, policy, or theory of justice. To hold that life and liberty are self-evident and inalienable rights is to substitute rhetoric for reality. Rhetoric may be an effective tool of political action but, unchecked by recognition of life's realities, it is a fatuous guide to policy.

Bioethics and Social Theory

Until midnineteenth century the long history of association between theories of life and theories of ethics had little practical effect upon political movements, policies, or social behavior. Organismic theories were explanatory rather than analytic. And more than a century following publication of Darwin's *On the Origin of Species by Means of Natural Selection* and the emergence of a science of genetics, no major political movement has been based upon or decisively influenced by the life sciences. True, evolutionary thinking had some influence on Karl Marx (who considered dedicating *Das Capital* to Darwin), and behavioral science was drawn into the service of political propaganda.[14] So-called social Darwinism for a short time influenced conservative politics, ostensibly providing a rationale for the social and economic inequalities associated with late nineteenth and early twentieth century industrial capitalism.

It was this aspect of Darwinism that incensed social scientists and ethicists. But it was a social Darwinism of the sociopolitical right that received their attention and initially received some support from sociologists; but with little following. *The Encyclopedia) the Social Sciences* describes social Darwinism

> as a short-lived theory of social evolution, vigorously discussed in America, which rationalized and justified the harsh facts of social stratification in an attempt to reconcile them with the prevailing ideology of equalitarianism. The emergence of social Darwinism was perhaps the most visible effect on the social sciences of Charles Darwin's *On the Origin of Species.*[15]

There was also a social Darwinism of the left, exemplified in writings by Friedrich Engels. But the ethical propositions that dominated twentieth century politics were not only largely uninfluenced by biological concepts, they were in some respects influenced negatively, that is in reactive opposition to bioscience. Biology became the subversive science. Welcomed by liberals and socialists when it undermined religious and political conservatism, it was conversely viewed with alarm when it seemed to question the conventional "truths" of democracy and socialism.

The era of Darwin and Wallace and the emergence of natural selection as an evolutionary hypothesis may be regarded as a watershed in bioethcial theory. Prior to the midnineteenth century, science afforded little substance for a defensible biology-based ethic. Pre-Darwinian ethics, insofar as the phenomena of life were invoked were based upon deductive analogy rather than upon empirical evidence. Following Darwin, there began a debate over the legitimacy of science as a source for ethics, a dialog that has continued into our own time. The debate has only in part (primarily in its earlier years) been concerned with the validity of the natural selection hypothesis. The main issue is no longer with the truth of science per se (even the Christian fundamentalist creationists invoke science (as they see it) as more valid than that accepted by most professional scientists). It is with the speculative implications of evolutionary science and genetics for human conduct and morality that dissention arises. The implications multiply with the advancement of biological knowledge and technique especially through discoveries in the mechanics of evolution—genetics—and the mediating influences of the environment, natural and artificial.

Assuming that politics implies, in some sense, an ethic, what is the source of ethics for a politics that seeks to shape its values and its goals with the assistance of biology? Does biology afford a sufficient basis for ethics? This question is conditional; its answer depends not only upon the criteria for *sufficiency*, but also upon the definition of *ethics*. We have noted a contemporary tendency to define ethics subjectively, saying in effect, ethics are what I regard as ethical—each person being his own judge of ethical conduct. An additional complication in relating bioethics to social theory is the *way* in which biological fact receives an ethical interpretation and the way in which that interpretation is applied to political theory, policy, and action. The social consequences of bioethics should not be prejudged. The defense of bioethics rests upon the proposition that ethical judgments based on biological ignorance are not likely to be as conducive to human happiness, well-being, or survival as ethics that take account of biological realities.

Because biocracy is concerned with the meaning of survival as a goal of politics and because there are wide differences of opinion over the ethics of survival as a goal, our concern with bioethics will be focused upon three closely related aspects of the survival issue. These aspects are first, the meaning of "fitness," second, homeostatic or equilibrium-seeking mechanisms in physiology and society, and third, criteria, if there are any, that discriminate between "right" and "wrong" in survival.

The Fitness Concept

The debate over fitness was not an immediate response to the evolutionary hypotheses of natural selection, struggle for existence, and survival. It began instead as a reaction to so-called social Darwinism, which was an extrapolation of the fitness doctrine to economic and social competition.

It was in reaction to this interpretation that Thomas Henry Huxley, Britain's leading expositor of Darwinian theory, took the occasion of the Romanes lecture at Oxford University in 1893 to contrast the evolution of nature which he called a cosmic process with the ethical process of human society. In an analysis and critique of Huxley's lecture, the American philosopher John Dewey in an address at the University of Chicago in the summer of 1897, summarized Huxley's argument (which was not Dewey's) as follows:

> The *rule* of the cosmic process is struggle and strife. The rule of the ethical process is sympathy and cooperation. The *end* of the cosmic process is the survival of the fittest; that of the ethical, the fitting of as many as possible to survive. Before the ethical tribunal the cosmic process stands condemned. The two processes are not only incompatible but even opposed to each other. Social progress means the checking of the cosmic process at every step and the substitution for it of another, which may be called the ethical process; the end of which is not the survival of those who happen to be the fittest in respect of the whole of the conditions which exist, but of those who are ethically the best. The practice of that which is ethically best—which we call goodness or virtue—involves a course of conduct which in all respects is opposed to that which leads to success in the cosmic struggle for existence. . . . The cosmic process has no sort of relation to moral ends. The imitation by man is inconsistent with the first principles of ethics. Let us understand once and for all that the ethical progress of society depends not on imitating the cosmic process, still less in running away from it, but in combating it.[16]

Huxley subsequently qualified this sharply dualistic concept, pointing out that man is indeed a part of nature and that the cosmic forces operate independently of his wishes. But he regarded nature, including human nature, as an unfinished product, susceptible to improvement through the application of human intelligence and morality. Yet in his anxiety to refute the exaggerated and misapplied concepts of fitness by so-called social Darwinists, Huxley missed what should have been his ultimate conclusion: that although cosmic nature did not provide a model for human behavior and that ethical principles were not deducible directly from nature, man could nevertheless learn from nature in endeavoring to improve upon it.

John Dewey found that this omission logically followed from Huxley's misinterpretation of the meaning of fitness. To Huxley's assertion that the test of social progress and ethics was " . . . not the survival of those who happen to be the fittest in respect of the whole of the conditions which exist. . . . " Dewey replied that "the conditions with respect to which the term 'fit' must *now* be used include the existing social structure with all the habits, demands and ideals found in it." And he therefore concluded that:

> the fittest with respect to the whole of the conditions is the best; that, indeed, the only standard we have of the best is the discovery of that which maintains these conditions in their integrity. The unfit is practically the anti-social.[17]

Elaborating on this theme Dewey further declared that:

> the environment is now distinctly a social one, and the content of the term "fit" has to be made with reference to social adaptation. Moreover, the environment in which we now live is a changing and progressive one. Every one must have his fitness judged by the whole, including the anticipated change; not merely by reference to the conditions of today, because these may be gone tomorrow. If one is fitted simply to the present, he is not fitted to survive. He is sure to go under. A part of his fitness will consist in that very flexibility which enables him to adjust himself without too much loss to sudden and unexpected changes in his surroundings. We have then no reason here to oppose the ethical process to the natural process.[18]

Thomas Henry Huxley's particularized and narrow view of ethics and of the struggle for existence unfortunately has tended to set the tone for much of the subsequent debate over the ethics of social Darwinism. Subsequent advances in our understanding of evolution and ethology, the science of animal behavior, have reinforced Dewey's understanding of fitness, not Huxley's interpretation. The "ape and tiger promptings" that Huxley regarded as immoral not only reflected the prevailing ignorance and misinterpretation of animal behavior, but was a misapplication of moral judgment to acts neither moral nor immoral. Dewey concluded his analysis by observing that although some theologians and moralists welcomed Huxley's reassertion of the dualism of the natural and ethical, he questioned:

> whether the spiritual life does not get its surest and most ample guarantees when it is learned that the laws and conditions of righteousness are implicated in the working processes of the universe; when it is found that man in his conscious struggles, in his doubts, temptations, and defeats, in his aspirations and successes, is moved on and buoyed up by the forces which have developed nature; and that in this moral struggle he acts not as a mere individual but as an organ in maintaining and carrying forward the universal process.[19]

Years later in his study on *Human Nature and Conduct: An Introduction to Social Psychology* (1922), John Dewey summed up his position on the relationship between science and ethics:

> A morals based upon the study of human nature instead of upon disregard for it would find the facts of man continuous with those of the rest of nature and would thereby ally ethics with physics and biology.[20]

In an address in 1945 before the section on History and Philosophy of Science of the American Association for the Advancement of Science, biomedical scientist Chauncey D. Leake continued Dewey's line of reasoning. Applying it to the ethics of survival, Leake argued that the methods of science could be applied to the formulation and testing of ethics and values. He declared that:

> with increasing knowledge derived from the various levels of scientific endeavor, it now appears that we have sufficient facts of a verifiable nature regarding ourselves and our environment to estimate the moral value of their implications; that is to determine their ethical significance.[21]

Leake did not propose to determine ethical significance through pure deductive reasoning, but through examination of the consequences of behavior. He concluded that: "Whatever is conducive toward the continued survival of that particular individual or that particular species is 'good' for it—in the very significant sense that if the species fails to survive, 'goodness' has no further meaning for that species." But in a world of interdependent interrelating organisms, goodness in the larger sense requires consideration of the factors conducive to the survival of all species. This leads further to consideration of the relationships among living things.

Emphasizing the significance of adaptation in survival and the importance of adjustment to environmental conditions as a factor of survival, Leake concluded that these circumstances applied equally to nature and to human society. He agreed with biologist Edwin Grant Conklin that "Biologically life is maintained by continual balance, cooperation, compromise, and the same principles apply to the life of society." And he cited the conclusion of zoologist Samuel J. Holmes that:

> Morality becomes just one phase of the adjustment of the organism to its conditions of existence. As a good body is one that runs smoothly and efficiently in the maintenance of its vital functions, so a good man is one whose conduct not only maintains his life on an efficient plane, but conduces to the enhancement of the life of his social group. Peoples may believe that their moral principles derive from a supernatural source, but one potent reason for their adoption is their conduciveness of survival.[22]

The practical or operant expression of this principle was symbiosis—mutually beneficial association—which, declared Leake, " . . . implies the culmination of the dynamic interaction of organism and environment in which the surviving organism tends to be best fitted to the environment surrounding it." Thus the analysis of fitness led to considerations of adaptive processes and relationships between organisms and their changing environments. This process of "ethicogenesis" as Leake called it, implied an evolution of ethics under the influence of man's growing knowledge of himself and his environment. Fitness or the capacity for survival was a normative test of goodness in this naturalistic concept of ethics. Symbiosis was, in principle, the condition most conducive to survival in a world of multiorganic complexity. But symbiosis implied stability and continuity in interorganic or interspecies relationships. Stability and continuity between organisms required a corresponding condition within organisms. In an ever-changing universe, human fitness must result from an adaptive process, which if survival is an ethical goal, must by definition also be an ethical process. This dynamic aspect of the fitness concept was identified by some biologists as a process of dynamic homeostasis.

The Ethic of Dynamic Homeostasis

The concept of homeostasis provided a more explicit and demonstrable linkage between biology and sociology than the more abstract ideas of organic analogy and fitness. It has become one of the many basic ideas underlying biocratic thinking, but should not be regarded as merely a transfer of physiological explanations to sociological theory. Homeostasis may perhaps be best described as an *effect* rather than a process. It may be understood as a *condition* resulting from the interaction of diverse processes. As with other biological phenomena, direct analogies with social behavior may be misleading. The concept of homeostasis does have affinities with organismic and equilibrium theories of society, but as we have noted (in chapter 3), its relationship to these theories is much qualified and requires careful examination.

The most explicit development of the idea of homeostasis as a basis for social ethics (previously noted in chapter 3) was by University of Chigago zoologist Alfred E. Emerson in a paper entitled, "Dynamic Homeostasis: A Unifying Principle in Organic, Social, and Ethical Evolution."[23] Emerson's thesis began with the proposition " . . . that intricate social behavior evolves and is an expansion of biological antecedents, and biology should be able to supply us with basic principles underlying social coordination." "The biologist," he said, " . . . will only assist in laying foundations upon which the social scientist can build."

It is difficult to state Emerson's thesis concisely. We may recall that homeostasis may be described as the effect of coordinated processes that tend to maintain steady states in organisms and at other levels of organization. Homeostatic processes are characteristically of a cybernetic or negative feedback character, producing a stabilizing or corrective response in relation to change in the environment. Emerson believed that increased homeostasis of living systems organized in various levels of integration seemed ". . . to be a general trend of biological and human individual and group evolution." He recognized exceptions to this trend and areas of life experience to which it did not apply, but he nevertheless saw in it a major factor for biological and social survival. The unfolding or refinement of ethical awareness occurred at the most advanced or complex levels of social integration and survival was necessary to this attainment. Thus homeostatic processes which maintained the integrity of organisms and societies must be regarded as "good." Hence the degree to which the behavior of individuals or societies could be described as ethical, related significantly to the effect of that behavior on physiological or societal homeostasis.

Closely related to this concept of homeostasis was the idea of an evolutionary progression from simpler to more complex levels of integration. This concept, developed especially by physiologist Ralph W. Gerard, provided a unified theory of organization linking inorganic matter to biology to society, a progressive relationship expressed teleologically by Pierre Teilhard de Chardin in his *Phenomenon of Man* (1961).

The concept of levels of integration became an important element of general systems theory to which Gerard was also a major contributor. The search for principles underlying systemic relationships and the evolution of organizational complexity moved the discourse on the relationship between biology and society more firmly into a mode anticipated by those commentators who saw social behavior evolving from a biological base. If systems theory could explain how certain organization-maintaining processes, such as dynamic homeostasis, operated at "higher" levels of life it should then be possible to understand why certain behaviors could be regarded as ethical. It should be possible to understand in what sense and with what qualifications life-sustaining behaviors and prcesses are "good," and why behavior that respects the ecological integrity of life should be regarded as ethical.

From this it follows that the philosophical dispute over the relevance of biological analogy to sociology and to ethics is no longer an appropriate or productive way to address interrelationships between nature and culture. Analogies, homologies, and comparisons generally afford clues to the possibility of common systems functions or relationships. The ethical significance of this more fundamental approach to the study of factors common to biological and sociological processes is that it leads toward understanding of why particular behavior may be regarded as ethical or unethical. It enables us to supplement or correct intuition with reason. It offers explanation, not mere assertion. This conclusion is hardly new, even though widely unpopular for nearly a century. As early as 1890 in *An Introduction to Social Philosophy,* John S. Mackenzie wrote: "It can hardly be denied that some of the [biological] analogies which recent writers (especially Spencer and Schäffle) have traced out are truly enlightening; but they help us, I think, chiefly insofar as they enable us to clear our minds of other analogies which are less adequate; such as those derived from chemistry or mechanics."[24]

Bioethics in Political Action

Against the foregoing background of concept and rhetoric we may now examine what practical difference a bioethical orientation makes in conflicts over political choice. A bioethical bias in values does not necessarily predetermine conclusions. Persons committed to bioethical principles may differ in the weighting and construction of particular facts. They may differ as to the extent to which a biology-based consideration offers the appropriate response to an ethical problem under a particular set of circumstances. Consequently in examining the ethical content of biopolitical issues our examples illustrate *a* bioethical approach and not *the* bioethical solution. But because of the way in which the world is put together, bioethical options are limited. There are not an indefinite number of ways to behave ethically or realistically in relation to the natural world.

Survival as a practical problem for bioethics is reduced to subsidiary problems of *what* or *who, how* or *when, where* and *why.* At these more particular levels of experience real personal and political choices are made. For these choices no ethical system, including the bioethical, offers the prospect of results universally acceptable to the persons, organisms, or systems affected. Conflict over the ethical basis of choice is therefore almost inevitable. Ethics to most people are *their* ethics. Conduct regarded as ethical by one criterion may be regarded as unethical by another.

It is not necessarily true that people regard as ethical only that which redounds to their personal advantage. Ethical beliefs have arisen in both Christianity and Hinduism, for example, that work against the material welfare and even survial of the believers. The ethics of martyrdom and of Karma have not worked to the survival of particular individuals; whether these ethics have in actuality had a survival value for gene pools, culture, or society is more difficult to determine. Sociobiological hypotheses regarding the genetic basis of altruism and inclusive fitness, however, suggest that if survival of the gene pool is benefitted by self-sacrifice, this form of "ethical" conduct may be biologically advantageous. The more common circumstance in which ethical principles work against survival is where their ultimate consequences are unperceived. Three examples will illustrate the effect of knowledge on ethics and subsequently upon political choice.

The first example pertains to the survival consequences of environmental depletion and degradation (see chapter 4). For centuries the great mass of humanity has seemed unaware of the cumulative effect of its activities upon the life-support capabilities of the environment. History, archaeology, anthropology, ecology, physiology, medicine, and the behavioral sciences have provided knowledge of the cause and effect sequences of exploiting the human environment, interior as well as exterior, beyond capacities for self-renewal. Information has permitted the identification of tolerances, thresholds, and limits beyond which survival may be jeopardized. Under a bioethics of survival it would be wrong to transgress these parameters regardless of whatever man-made social, legal, or religious "rights" may be claimed. Thus if the ultimate survival of humanity depends, or even *may* depend, upon the continuing viability of land, water, and air, human behavior that impairs their life-sustaining properties is unethical.

The ethics of biocracy would therefore deny any alleged "right" that would impair planetary life-support capabilities. Thus an economic judgment regarding the present "best" use of a piece of land would be acceptable as best under biocracy only if that use was consistent with maintenance of the life-support system. Clearly, biocracy would restrict options and slow some decision processes. But the occasions upon which it would do so would be those in which the proposed action was, in the balance, prejudicial to human welfare for reasons that could be specified by recourse to verified (i.e., scientific) knowledge.

A second example pertains to the respective responsibilities of the individual and society in relation to human health. The state of ethics in modern society appears in total confusion over who is responsible for human health, and how this responsibility should be given practical expression. In most ethical systems it appears that responsibility for the care and treatment of the human body is divided. Responsibilities are characteristically allocated to individuals (usually adults), to families, and to the larger society. Does a person have a right to live in a manner that impairs his health to the extent of infringing upon the rights of others and of burdening their lives and possibly jeopardizing their survival? Here again science provides a forecast of consequences which alters the basis of ethical choice. The findings of physiology and behavioral science do not support a right to the pursuit of happiness through the consumption of food, alcohol, tobacco, or more dangerously addictive drugs to an extent that impairs the health of the consumers. The argument that how much I eat, smoke, or drink is nobody's business but my own has no standing whatever under biocracy. Individual conduct becomes a potential public concern at whatever point the health, welfare, and convenience of others are threatened. Behavior that raises taxes and insurance rates, preempts the services of physicians and hospitals, adds to the work load and worries of others is not bioethically defensible.

A third example relates to ethnic survival. Here is one of the more difficult conjunctions of ethics and politics. History and anthropology may have nothing to say regarding which societies or cultures "should" or will survive, but they offer clues to the reasons that some have survived and most have not. The definition of what constitutes a particular culture, society, or ethnic group is a matter of considerable ambiguity, and to an extent all classifications are arbitrary. Yet it may not be necessary to resolve this problem of group identity in relation to the survival issue, although the way in which groups are defined may affect their survival opportunities.

The essential condition for ethnic survival is the maintenance and transmission of ethnic integrity. Beyond a point which may be difficult to ascertain, cultural and genetic dilution threaten the integrity of specific ethnic groups. For example, in modern national states, restriction of immigration through quotas designed to maintain ethnic compatibility are bioethically correct if the continuity and survival of the traditional culture is a consideration. Genetic change in a culture, whether described as dilution or enrichment, may or may not affect its prospects for survival.

Alfred Emerson's examination of the ethical implications of homeostasis led him to conclude that: "In the evolution of any social system we must expect social variability, a degree of social isolation, and selective assortment of the most efficient social characteristics producing homeostasis."[25] He recognized that in society as in nature, variation may be deleterious in effect. But he held that " . . . any social tendency drastically to restrict variation in ideas and actions may result in an unprogressive stabilization of the system." Nevertheless he concluded that: "Progressive social evolution is in part dependent upon partial isolation" and that "Either extreme isolation or extreme interchange would slow adaptive evolution."

But Emerson's analysis demonstrates how easy it is to slip from the scientific into the humanistic mode of reasoning. Influenced possibly by a theory of intraspecific inhibition against lethal violence, he found that " . . . the concept of the 'brotherhood' of all mankind rests upon firm biological principles. Ethics leading to firmer integration and mutual benefit between races of man is in conformity to biological trends." This conclusion is hardly helpful to the making of survival-conscious political choices. Emerson did not explain how the desiderata of racial integration and social isolation could simultaneously be realized, or consider the possibility that factors influencing aggression within species might differ with the species, and that man might represent a difference amounting to an exception.

The "brotherhood of man" is a humanistic rhetorical expression, suggestive of ethical relationships, but of no evident significance for a politics of survival. It is scientifically true that all mankind shares a common evolutionary genetic heritage. And this heritage is even more widely, although remotely, shared with all life on earth. The biochemical molecular biology of life is the same for all living species. But brotherhood is a specific familial relationship and to extrapolate the concept to society or humanity is not science but poetry.

The contribution of biology to the ethics of ethnic relationships is surely more modest, but more complex than the "brotherhood" metaphor. It is doubtful that biology will ever provide an ethical rationale for all biology-related human action. The human mind can conceive of circumstances and relationships that would almost universally be regarded as "good" that are not realizable in the realities of this world. Many of life's choices are unavoidably tragic—there being no "right" or "ethical" alternatives. Respect for ethnic diversity may be regarded as biologically neutral, but practical efforts to safeguard particular cultures and their ethnic integrity are likely to be attacked as socially unethical if devisive of more inclusive sociopolitical aggregations.

The decision of the Philippine government to protect the integrity of an alleged late-discovered stone age people on the island of Mindanao may be regarded as responsive to a growing biocultural ethic. But this "protection" could not prevent the intrusion of the modern world and the rapid acculturation of these people into early stages of modernity. There appears to be no way in which the European conquest and settlement of the Americas could be bioethically regarded as other than wholly unethical, but colonizers and missionaries saw the process very differently.

The ultimate limitation of biology to shape the ethics of political choice lies in the cultural origin of ethics. Nature per se is ethically neutral. Ethics having evolved with culture, its norms are those of culture, even when modified or corrected by scientific knowledge. Man is a culture-making animal, phylogenetically adapted to conceiving and maintaining cultural institutions and relationships. Emotional allegiance to cultural values has a survival value for that culture and the biological bearers of that culture. The enemies of cultures are more often the people of other cultures than the elements of nature. And these cultures characteristically embody the most lofty and compelling ideals of which their people at any given time are capable.

Thus we see the long history of holy causes and inalienable rights for which men have enthusiastically slaughtered or supressed their fellows. Whether there are genetic, physiologic, or biochemical factors in human aggression is as yet debatable. Is genetic omission or provision a factor? Studies of mankind's nearest relatives among the primate apes suggest that human aggression including warlike behavior is not unique to the human species. In his study *On Aggression,* Konrad Lorenz sums up the tragic ethical paradox of the human condition:

> . . . the need to control, by wise rational responsibility, all our emotional allegiances to cultural values is as great as, if not greater than, the necessity to keep in check our other instincts. None of them can ever have such devastating effects as unbridled militant enthusiasm when it infects great masses and overrides all other consideration by its single-mindedness and its specious nobility. It is not enthusiasm in itself that is in any way noble, but humanity's great goals which it can be called upon to defend. That indeed is the Janus head of man: The only being capable of dedicating himself to the very highest moral and ethical values requires for this purpose a phylogenetically adapted mechanism of behavior whose animal properties bring with them the danger that he will kill his brother, convinced that he is doing so in the interests of these very same high values. *Ecce homo!*[26]

Yet biobehavioral science may offer some hope of ethical advancement toward resolving this paradox. The need to control emotional allegiances by wise rational responsibility can be served by better understanding of the mainsprings of behavior, and of the things that may be done (or should not be done) to diminish the prospect of evil in the name of righteousness. Thus bioethics may not show us the way to make all tragic choices ethical. Advancement in biological understanding and technology may someday confront humans with some very fundamental choices that are profoundly tragic and from which there is no evasion. But it may reduce the number of such choices where through knowledge they may be avoided, and may mitigate the worst effects of those choices that must be made. The route toward the goal of human behavior that most men have regarded as ethical may be though neurology, biochemistry, and genetics. Philosophy may set the goals but bioscience may be necessary to achieve them.

Notes

1. On the subject of bioethics, see Warren T. Reich, ed., *The Encyclopedia of Bioethics*, 4 vols., (New York: Free Press, 1978); LeRoy Walters, ed., *Bibliography of Bioethics* (Detroit: Gale Research, 1975); F. J. Ebling., ed., *Biology and Ethics* (New York: Institute of Biology, 1969); Bernard Rensch, *Biophilosophy*, trans. C. A. M. Sym (New York: Columbia University Press, 1971); William B. Bondeson, et al., eds., *New Knowledge in the Biomedical Sciences: Some Moral Implications of Its Acquisition, Possession, and Use* (Dordrecht, Holland: D. Reidel, 1982); Elsie L. Bandman and Bertram Bandman, eds., *Bioethics and Human Rights* (Boston: Little, Brown and Co., 1978); Kenneth Vaux, *Biomedical Ethics: Morality For the New Medicine* (New York: Harper and Row, 1974); H. Tristram Engelhardt, Jr., *The Foundations of Bioethics* (New York: Oxford University Press, 1986); and Coleen D. Clements, "The Bureau of Bioethics: Form without Content Is Meaningless," *Perspectives in Biology and Medicine* 27 (Winter 1984): 171–82. See also *Bioethics Quarterly* (New York: Human Sciences Press, 1979–present). The Hastings Center, Institute of Society, Ethics, and the Life Sciences, 360 Broadway, Hastings-on-Hudson, New York, 10706, is an indispensable source of information on publications relating to bioethics; see especially *Bibliography of Society, Ethics, and the Life Sciences*, 1973–1978, retitled *Bibliography of Ethics, Biomedicine, and Professional Responsibility*, 1979–present. The extent of published work is large and growing. Bioethical articles appear frequently in the journals *Environmental Ethics: Philosophical and Policy Perspectives; Perspectives in Biology and Medicine; Science and Religion;* and *Zygon.*

2. Alasdair McIntyre, *After Virtue: A Study in Moral Theory* (Notre Dame, Indiana: University of Notre Dame Press, 1981), 8.

3. Van Rensselaer Potter, *Bioethics: Bridge to the Future* (Englewood Cliffs, New Jersey: Prentice-Hall, 1971).

4. Alexander Pope, *An Essay on Man, Epistle I, On the Nature and State of Man with Respect to the Universe.*

5. Cf. Arthur O. Lovejoy, *The Great Chain of Being: A Study of the History of an Idea* (Cambridge: Harvard University Press, 1936).

6. Potter, *Bioethics.* See also Robert H. Blank, "Bioethical Decisions: The Political Context and Challenges," *Bioethics* 3 (Fall–Winter 1981): 163–79.

7. Charles Darwin, "An Historical Sketch of the Progress of Opinion on the Origin of Species Previously to Publication of the First Edition of This Work," *On the Origin of Species by Means of Natural Selection*, any edition.

8. Ibid.

9. For earlier attitudes toward eugenics, see Horatio Hackett Newman, *Evolution, Genetics, and Eugenics*, 2d ed. (Chicago: University of Chicago Press, 1925); Donald K. Pickens, *Eugenics and the Progressives* (Nashville, Tennessee: Vanderbilt University Press, 1968); and Loren Graham, "Science and Values: Eugenics Movements in Germany and Russia in 1920s," *American Historical Review* 82 (December 1977): 1133–64. Writings on eugenics characteristically show ideological biases, pro and con. See Mark H. Haller, *Eugenics: Hereditarian Attitudes in American Thought* (New Brunswick, New Jersey: Rutgers University Press, 1963); and Daniel J. Kevles, *In the Name of Eugenics: Genetics and the Uses of Human Heredity* (New York: Alfred A. Knopf, 1985).

10. Alfred E. Emerson, "Dynamic Homeostasis: A Unifying Principle in Organic, Social, and Ethical Evolution," *Scientific Monthly* 18 (February 1954): 67–85.

11. Cf. James Miller, *History and Human Existence from Marx to Merleau-Pon* (Berkeley and Los Angeles: University of California Press, 1979), 108.

12. Ibid., 106.

13. The literature is multidisciplinary and very large. Examples of particular relevance here include Robert C. Barker, *Ecological Psychology: Concepts and Methods of Studying the Environment of Human Behavior* (Stanford, California: Stanford University Press, 1968); Aristide H. Esser, ed., *Behavior and Environment: The Use of Space by Animals and Men* (New York: Plenum, 1973); and David Lowenthal, ed., *Environmental Perception and Behavior* (Chicago: University of Chicago, Department of Geography, Research Paper 109, 1967).

14. Miller, *History and Human Existence*, 107.

15. Sol Fox and Larry S. Krucoff in *International Encyclopedia of the Social Sciences* (New York: Macmillan, 1968), 402. Nevertheless, the authors conclude that "even if biological Darwinism is no longer of direct relevance for the social sciences, its heritage includes some very basic concepts and methods" (405).

16. See John Dewey, "Evolution and Ethics," *Scientific Monthly* 78 (February 1954): 58. This essay was reprinted from *The Monist* 8 (1897–1901).

17. Ibid., 60.

18. Ibid., 60–61.

19. Ibid., 66.

20. John Dewey, *Human Nature and Conduct: An Introduction to Social Psychology* (New York: Henry Holt, 1922), 12.

21. Chauncey D. Leake, "Ethicogenesis," *Scientific Monthly* 60 (April 1945): 245–53. For another evolutionary view of ethics see Hudson Hoagland, "Ethology and Ethics: The Biology of Right and Wrong," *Zygon* 2 (March 1967): 43–58.

22. Samuel J. Holmes, "Darwinian Ethics and Its Practical Application," *Science* 90 (11 August 1939): 117. Holmes was Professor of Zoology at the University of California.

23. Alfred E. Emerson, "Dynamic Homeostasis: A Unifying Principle in Organic, Social, and Ethical Evolution," *Scientific Monthly* 78 (February 1954): 67–85.

24. John S. Mackenzie in *An Introduction to Social Philosophy* (Glasgow, Scotland: James Maclehose, 1980).

25. Emerson, "Dynamic Homeostasis."

26. Konrad Lorenz, *On Agression,* trans. Marjory Kerr Wilson (New York: Harcourt, Brace, and World, 1963), 274.

7 Biocracy and Bureaucracy: Coping with Conflict

This book advances the proposition that growing knowledge of the biology of human behavior threatens certain assumptions upon which modern government, law, and bureaucracy have been based. Some observers would declare that the threat is already present and there is increasing evidence to give the argument plausibility. The theories of human behavior that underlie modern government have been deduced from prescientific premises; governmental institutions, laws, and programs based upon these theories are thus founded upon propositions that may be true but are not demonstrably valid. Their tenets have remained acceptable because under the circumstances of the past they seemed to have worked to satisfy social needs and purposes. But should they fail to meet the challenge of new conditions, fail to provide reliable answers to persisting social questions, and to retain credibility in the light of scientific inquiry might not, in time, their validity as foundations of our sociopolitical order be called into question?

Biology and the new science of sociobiology now force lawmakers, administrators, and judges to consider issues that were once beyond the limits of political concern.[1] Biological findings continue to unsettle assumptions long accepted in public law and policy. Even the necessity and durability of bureaucratic centralization is being questioned as possibly contrary to human nature as revealed by biobehavioral science. In effect, biology and bureaucracy are approaching a confrontation comparable to the philosophical conflict between physical science and the medieval church.

One need not look far to discover that political "truth" may not be scientifically true. Law and the rhetoric of government are filled with

---

Based in part on "Biology and Bureaucracy: The Coming Confrontation," *Public Administrative Review* 40 (January–February 1980): 1–12.

assertions that cannot be proved and, indeed, often can be shown to be false. This situation may create political difficulties when a scientific proposition, valid under some conditions, is invalid under others. Scientific conclusions tend to be of this sort, and are not always compatible with laws and doctrines that purport to state absolute principles, or are phrased in language that accommodates no exceptions. Is it possible that the all too numerous failures of government today may in part be explained by reliance upon flawed premises and inappropriate criteria for testing alleged truth? Can "right" answers be found to problems which have been erroneously defined and which assume relationships and probabilities that may not exist? The premises and definitions of eighteenth and nineteenth century political theorists are now being tested by science and experience, and it is far from certain that they can pass all the tests.

At the moment, defenders of modern bureaucracy may have presumption in their favor when they ask: "How do you know that our behavioral theories are in error or that our policies do not address what you call the *real* problems? Until they are proved wrong, are we not entitled to assume that they may be right or in any case the best available?" But such defenders may be asked to explain why there is so little confidence in the ability of governments to cope with society's problems. Why do so many people regard the problem-solving record of modern government to be (with few notable exceptions) impressive for its failures? And why have historical bureaucracies, intended to strengthen social planning and control, so often become obstructive burdens to social vitality, swelling to inordinate size during the declining years of political systems? Does recourse to bureaucratic government follow from a misreading of human nature and the conditions conducive to human cooperation?

The most notable problem-solving achievement by contemporary government was the American Apollo program—which placed men on the moon and returned them to earth. Its problems were preeminently technical but, to the extent that behavioral problems were encountered, Apollo drew upon biopsychology and physiology—not upon social theory. As an organizational effort it was complex but singularly nonbureaucratic, involving careful coordination of many separate governmental and nongovernmental subdivisions through a matrix of contractual relationships. The subsequent midair explosion of the space shuttle Columbia has in retrospect been seen as an imposition of certain human values on the space program that were inconsistent with the scientific requirements for safety.[2]

Today, the horizon of our public life is clouded as one political storm after another sweeps toward us. Accompanying these disturbances, and shaking confidence in established political doctrine, come cold blasts of new scientific knowledge regarding man's inner nature. Even the early effects of this knowledge have been destructive to old assurances, shattering traditional assumptions regarding human motivation and behavior and leaving society, for the present, without substance upon which new social policies can be securely based.

If only by intimation, the impact of bioscience and the new science of sociobiology is beginning to be felt in those public services most directly concerned with behavior—in mental health, medicine, domestic relations, housing, urban planning, criminal justice, education, and welfare. The new biology is often perceived to have an equivocal or negative effect upon established programs and practices that makes it unwelcome. But, a positive impact might be even more unwelcome. The present state of social biology suggests that many of our basic assumptions regarding intelligence, altruism, criminality, learning, and dependency are not supported by scientific evidence. This in itself does not prove them wrong, but if our assumptions are not only arbitrary but substantially in error, policies based upon them could defeat their well-intentioned purposes. How much more disturbing might it be if strong evidence derived from sociobiological theory and research were to contradict fundamental tenets of liberalism and democracy? What perplexities might follow persuasive evidence that large bureaucratic hierarchies were maladapted to human creativity and responsibility? Can sociobiologists be characterized as miners and sappers undercutting the foundation of a free and orderly society?

A New Heresy?

Biologically based social theories are conceptual nuclei around which a new sociopolitical heresy appears to be forming. This could be heresy in the true sense of the term. For the present it may have little visible effect upon the ordinary business of politics and government. No sign of apprehension can be detected in the United States Congress or major political parties anywhere. But beliefs and institutions persist long after confidence in them has become weakened. They tend to persist until they encounter conflict with which they cannot cope.

Some findings in biology and sociobiology suggest theories of human nature and behavior that conflict with traditional basic beliefs; and implications derived from this evidence could catalyze new concepts and doctrines of society and government. It is conceivable, even probable, that from this conceptual evolution, and in response to social events, new political ideologies might emerge. Evidence to date is no more than suggestive, but what it suggests does not support the survival prospects of totalitarian or authoritarian political systems. Neither does it confirm some of the basic assumptions of mass democracy, socialism, and the liberal welfare state. If the beliefs of humanity about its own basic nature have been at variance with reality, it follows that the political theories of the "left" and the "right" may be equally, although differently, wrong. And it may also follow that an apparent human tendency to develop large, impersonal, centralized bureaucracies to cope with social needs could be a monumental error compounded by man's misunderstanding of his innate needs and limitations. To discover the error, if in fact it can be demonstrated, does not, however, automatically supply a reliable corrective.

This new disturbing body of thought called sociobiology appears heretical because, by implication, it disputes and even challenges prevailing orthodoxies—civic, academic, and religious.[3] Sociobiology offers no support for unqualified and simplistic propositions that men are born free and equal, or that they have inherent and inalienable rights. Yet it does suggest by implication the kinds of freedom possible in a world in which not all things are possible. And it illuminates the ways in which equality might be meaningful in a world of variety and diversity. As social concepts move from their intuitive and rhetorical character in the past toward a scientific future, their meaning seems likely to change, but their essentially moral content is not necessarily lost. Yet in discovering more fundamental meanings for freedom and equality, sociobiology may deprive these terms of much of their rhetorical appeal. As for rights, sociobiology does not support conclusions that private property or a living wage are *natural* rights, or that the Marxist proletariat is the universal class entitled to inherit the future. In effect it desanctifies the political scriptures of liberal democracy and socialism, but without giving support to doctrines that would suppress one part of humanity for the benefit of another.

By implication, sociobiology opens to question the present structure of the social and behavioral sciences. Sociobiology is a practical refutation of the proposition almost universally accepted by recent generations of social scientists that biology is biology and sociology is sociology—and seldom the twain should meet. To immunize students against the errors of social Darwinism and organismic theories of the state, several generations have been taught that analogies between biology and sociology are inherently false. This generalization was extrapolated from the errors of particular analogies. And it is still widely believed that animal behavior has no relevance for human behavior, although field and clinical studies of primates and other animal species are making this opinion harder to defend.

At the moment, sociobiology has limited persuasiveness. Its propositions are not widely known and may not be generally accepted even among biologists. Yet sociobiology may claim a scientific status to the extent that its propositions can be subjected to tests of falsification. Of course, because a proposition cannot be proved false does not invariably affect its position in relation either to doubt or to faith. Revealed truth has an advantage when it is not accessible to scientific tests for falsification. Some people prefer to believe that there are inalienable natural rights of man, just as some people preferred to believe that there were divine rights of kings; and evidence intended to demonstrate the falsity of a belief does not automatically change the opinions of believers. But when scientific evidence gains strength, unsupported opinion tends increasingly to be forced to the defensive.

In the long run, ideas that "work," in the sense that they increase the prospects for desired survival, have a marginal advantage over those that do not. Imperfect truth may not always serve success or survival better than does persuasive falsehood, but, over time, statistical probabilities are in its favor. Heresy is outside the pale of dominant belief—to be taken seriously when it is perceived to threaten that dominance. The character of some attacks upon sociobiology by socialist-liberal activists suggests a perception of it as heretical. It is being taken seriously. Its implications are uncertain but suggestive and the possibility of its more complete and substantiated development adds to the fears of some critics. Sociobiology is some distance from catalyzing a theory of social action. But how great a distance in time—years or decades or perhaps never?

What Sociobiologists Say

The concept of sociobiology had been around for some time before publication of E. O. Wilson's *New Synthesis* (1975) stimulated an upsurge of public attention.[4] With respect to its relationship to biocracy and the study of biopolitics, two distinctions must be kept in mind. First, the study of sociobiology is very different from the conventional study of biological influences on society or the studies called biopolitics. Sociobiology represents a more focused and coherent singular approach to its subject matter. Biological influences, as considered in this book, embrace a greater number of impacts described in a wide range of biological disciplines, for example in ecology, microbiology, medicine, agriculture, and nutrition. Sociobiology, selectively incorporates some elements from all the social and behavioral sciences and substantial parts of the humanities while omitting others. Second, sociobiology defined as study of the biology of social behavior, is not merely the aggregate findings of investigators called sociobiologists. The concept that social behavior is significantly and even definitively influenced by biological factors is a hypothesis independent of the work of any particular group of investigators.

The more immediate challenge to contemporary political assumptions will probably not come from sociobiology narrowly defined. Findings in genetics, biopsychology, and biomedical research have already had a noticeable effect upon decisionmakers in government and scholars in academia. For example, opposition to investigating the genetics of intelligence or of sexual behavior may, in part, be explained by fear that findings might place some basic religious or democratic assumptions on the defensive, or that present theories of criminal justice might be compromised should it be established that a significant amount of sociopathic behavior was genetically influenced. The question, "what now?" would surely arise. Because sociobiologists have made the most explicit attempt to explain social behavior in genetic terms, their propositions offer the most direct challenge to conventional social theory. But far more important than the explicitness of the linkage between biology and sociology is the fundamental nature of the challenge. Here is how a leading sociobiologist, E. O. Wilson, puts the issue:

> Biology has traditionally affected the social sciences only indirectly through technological manifestations, such as the benefits of medicine, the mixed blessings of gene splicing and other techniques of genetics, and the specter of population growth. Although of great practical importance, these matters are trivial with reference to the conceptual foundation of the social sciences. The conventional treatments of "social biology" and "social issues of biology" in our colleges and universities present some formidable intellectual challenges, but they are not addressed to the core of social theory. This core is the deep structure of human nature, an essentially biological phenomenon that is also the primary focus of the humanities.

"Biology," writes Wilson, "is the key to human nature, and social scientists cannot afford to ignore its rapidly tightening principles."[5] Sociobiology has been defined by David Barash as "the systematic study of the biological basis of all forms of social behavior in all kinds of organisms. . . . "[6] And because its scope extends to all forms of society it should, in theory, be able to ascertain in what respects human social behavior is unique, and to what extent basic biological drives are overlayed or rationalized by human culture.

With regard to its political implications, Roger D. Masters has identified three critical concepts of sociobiology as: "reproductive success, inclusive fitness, and evolutionarily stable strategy."[7] It is essential to understand that the definitive unit for these concepts is the *species* and not the *individual.* Within the species, genetically differentiated and identifiable groups may be studied. But unlike moral philosophy, in the humanities and a large part of social and behavioral science the individual per se is not the primary object of inquiry. Of course, the findings of biology regarding the behavior of a species, or of closely related genetic subgroups, may illuminate the behavior of individuals. To the extent that basic universal biological forces are expressed through culture, sociobiology may help to close the traditional gap between humanistic and scientific explanations of human behavior. Nevertheless, the universal, transgenerational, and culture-neutral context of sociobiology puts it out of phase with nearly all conventional social and political theory.

The sociobiological concept of "reproductive success" may be described as the extent to which an organism (or genetically related group) contributes its genes to succeeding generations of the species. As a scientific concept, the term is meaningful only in relation to populations; to groups rather than to individuals. It is the consequence of inherent patterns of behavior, which may include the capacity for choice, that results in the survival of successive generations from a common source of genes. Not all genetically similar individuals need to reproduce to improve the reproductive success of their group or species. Individuals who do not pass on their genes may enhance the reproductive success of their genetically related group by supportive and protective behavior. From these hypotheses it is not difficult to generate propositions affecting family formulation, birth control, child rearing, and education. Thus, reproductive success in human society has cultural dimensions that may be influenced by the substrate of biologically conditioned behavior. And politicians and social scientists even before Malthus have recognized that breeding behaviors and strategies affect political futures.[8]

The interest of politicians in population dynamics relates more often to ethnicity and ideology than to genetics. Nevertheless, a connecting relationship between biology and politics has been recognized that is at least as old as the Hindu caste system and as recent as American immigration policy. At present, sociobiologists have little to say about linkages between genes and culture that lawmakers and administrators would find useful. But political leaders might consider the proposition that, in the long run, maximizing the number of offspring is not necessarily a reliable measure of reproductive success. As Roger D. Masters explains:

> If an investment of energy and resources is required to raise each offspring to maturity, unlimited increases in the number of children can be self-defeating. The population explosion of this century reminds us that

whenever resources are limited, the probability of survival can sometimes be increased by reducing the number of offspring.[9]

Fact that this may be, it is widely perceived as irrational in many prescientific societies. Thus from the perspective of sociobiology, the long-prevailing belief that political strength is enhanced by birth rates and increases in numbers is not axiomatic. But a belief so widely accepted must have some foundation in experience. In a society of high technology, Masters' observation may be right. Historically, however, nations capable of deploying large armies, and political parties able to mobilize mass votes, appear to have enjoyed a competitive advantage. And if social justice is construed to mean that all people should share equally in the opportunities of this world in direct proportion to their numbers, a breeding strategy enhancing individual fitness may defeat itself. In a "commons" situation all may eventually lose but, in the short run, the prolific breeders may have the advantage.[10]

As with reproductive success, "inclusive fitness" relates to collective rather than to individual characteristics. Inclusive fitness may be defined by what it measures: "net genetic representation in succeeding generations, including other relatives in addition to offspring."[11] Thus it differs fundamentally from social Darwinian interpretations of "survival of the fittest" as justification for rugged individualism. On the contrary, in most environments, a ruthless self-seeking life-style probably reduces inclusive fitness and hence reproductive success. There are, however, aspects of inclusive fitness which, to the extent that they reflect the biological substrate, have significant implications for politics and bureaucracy. For example, sociobiology indicates that altruism tends to vary directly with the degree of genetic relatedness—that is, altruism (other things being equal) tends to be strongest in relation to nearest of kin and weakest with respect to foreigners. This may explain why rhetorical concepts such as "brotherhood of man" and "social responsibility" have an uphill struggle when one tries to translate ideals into behavior. An implication following logically is that political and bureaucratic war against nepotism will probably be unending. In one form or another the tendency for parents to favor their children and for families to promote the interests of their members appears to be universal. Even in the most explicitly equalitarian human societies, people seem to find ways of giving their nearest kin a competitive advantage.[12]

As defined by biologists, notably by John Maynard Smith, "evolutionarily stable strategy" is behavior that tends to optimize reproductive success and minimizes the chances of extinction because of environmental change.[13] Obviously for genes and most animals, "strategy" in this sense does not imply conscious determination. Perhaps the clearest way to explain the concept is to define strategy as a pattern of behavior that has "worked" toward the survival of the species and probably will provide future survival advantage. Those animals whose number of offspring varies with food supply from year to year have incorporated an evolutionarily stable strategy into their reproductive behavior. The evolutionarily stable strategy, however, applies to behavior broadly (not only to reproduction), and with important modifications resembles mathematical games wherein each contestant acts to minimize his losses on the assumption that his opponent acts to maximize them.

But a minimax strategy based upon calculated reasoning is not wholly deducible from genetically conditioned behavior. Consciously adopted human strategies appear to rely upon a broad and often complex range of behavioral tendencies which, like the propensity of humans to develop language or to avoid incest, may be biologically derived but culturally conditioned. In the long run, innate behavior patterns in which culture is compatible with biology tend to maintain the stability of a species in a changing world and also to enhance its chances for survival.

It may be argued that the findings of biology in general, and of sociobiology in particular, are of minimal relevance to human behavior inasmuch as mankind has overcome its innate animality through cultural institutions: through religion, custom, law, and most of all, through government. Yet the gap between professed ideals and performance is a commonplace of human experience. This gap, acknowledged in the saying "the spirit is willing but the flesh is weak," equally expresses the dichotomy of culture (ideals) and biology (body). The psychobiological costs of cultural ideals have often proved to be too high for their realization. With some aid from sociobiological analysis, political reformers and public administrators might improve the prospects of some of their proposals and the probabilities of achieving desired results while avoiding undesired consequences.

Of course some political objectives may run counter to biological tolerances and induce counteractive responses, but these consequences may not occur until after the political life time of the perpetrators of the errors. At the least, sociobiology provides a "cost-benefit" approach to the analysis of social behavior. It offers a way to discover the nature and range of the costs as well as the benefits accruing from particular forms of social behavior. To accomplish this, however, sociobiology must be able to specify and scientifically to establish criteria that will distinguish costs from benefits, or explain why the distinction cannot be drawn. The common failure of cost-benefit analysis as a tool of policymaking has been in failure to fully assess all real costs and benefits. Cost-benefit analogies in relation to sociobiology must be used with great caution. With respect to human society, the approach could have a limited but critically important applicability in identifying the more optimal strategies for the survival of societies under various circumstances. It is too early to fully assess the depth or extent of sociobiology; but not too early to recognize that it will have significance, and that developments elsewhere in biology and behavioral science should take it into account.

From Biology to Biopolitics

Commenting on the impact of "The New Sociobiology" a distinguished student of public affairs, Luther Gulick, has declared that:

> The full impact on public administration and the social sciences of these new illuminations in social biology is still to be felt. But already it is clear that we have a lot of new thinking and experimenting before us. Already we know that we are dealing with a new vision of man, and we are beginning to see that we must also deal with entirely new concepts of social action, social morality and social structures.[14]

Clearly, not all aspects of biological influences on public administration are new; moreover, social biology in the general sense, including all society-related aspects of biology, should be distinguished from sociobiology as a particular scientific proposition. Biology manifest through behavior has been inherent in politics ever since politics began; and biopolitics has now become a subfield of political science. A large part of law and custom has concerned such biology-related matters as sexual customs, racial relations, marriage, inheritance, childbearing, child rearing, treatment of illness and death, animal husbandry, agriculture, and many more. But sociobiology as science—as search for verifiable knowledge of the deep causes of social behavior—has initiated a wholly different relationship with politics.

The *traditional* relationship was that various biological aspects of life, such as mating arrangements and family formation, were defined and regulated by social consensus, often through government. Biological behavior was classified and interpreted culturally; and the basis of political judgment was usually declared to be moral or rational—neither of which were believed to be rooted in human biology. On the contrary, their origins were believed to be spiritual or superorganic, transcending the biological (i.e., animal) nature of man. Judgments that today might be called biopolitical were not regarded as depending upon science in any respect. They were not scientifically informed judgments, but were sometimes intuitively correct.

A threat to the status quo occurs when, in an issue of active political controversy, a biological finding appears to make more sense than a political premise. But the contradiction does not necessarily clarify or sharpen the controversy as formulated in the political arena. A biological fact *may* indicate that an accepted social proposition cannot be true, and yet fail to provide an alternative truth. In some cases, such as beliefs regarding insanity or the causes of disease, a successful contradiction of public opinion opened the way to advances in biomedical research that ultimately altered public policies. In other cases, exemplified by the meaning of intelligence and the basis of personality, the social issue has been defined politically in language and through concepts that resist scrutiny by scientific methods. And so in these instances, biologically based argument leads to controversy and confused confrontation.

The right to life antiabortion issue is a case in point. The basic problem here, although not the one that has aroused controversy, is how the issue should be formulated. The right to terminate fetal life is the practical issue confronting legislature, courts, and administrative agencies; but for science to assist resolution of the issue, more fundamental matters must be addressed. These are: (1) how should *life* be defined? (2) do all organisms have a *right* to life? (3) at what stage of biological development does organic life become human personality, and (4) who, if anyone, should control human reproduction? The distinction between the specific issue of the legality of abortion and the definition of humanness is fundamental to understanding the confrontation between biology and politics. The former disturbs the structure of the political order, whereas the latter unsettles its foundations.

Until now, most biopolitical issues have been manageable within the normal structure of the political order. Controversies over such matters as the fluoridation of drinking water, compulsory vaccination, pasteurization of milk, smoking of cigarettes, use of pesticides, and sex education have been handled by legislators and administrators as normal political issues, hardly more touchy than taxation. However, in none of these controversies has politics or the bureaucracy become a forum for the reexamination of basic assumptions regarding human nature. Sociobiology, by contrast, makes explicit the need for such a reexamination. Government and its political institutions are probably not the right forum for reexamining such potentially devisive problems. Nongovernmental private but civic groups, churches, and universities appear to be the more promising places to work toward consensus. Government may then implement, as needed, after the social issues have been resolved.

The deliberate terminating of nascent life through abortion or infanticide provides illustration of how, under some circumstances, sociobiological tendencies can in the minds of some people "legitimize" behaviors otherwise regarded as repugnant. In the past, uncertain prospects for survival seem to have induced these forms of population control. Infanticide is almost totally rejected in civilized society today, but Mildred Dyckemann concludes that female infanticide among high caste Hindus in traditional India worked to increase their inclusive fitness.[15] However regrettable, it was not regarded as unnatural or immoral behavior. And Betty Meggers observes that infanticide of twins among Amazonian tribes was ecologically beneficial in relation to the absolute necessity for restricting the number of mouths to be fed.[16] In effect, the Amazonians were practicing an evolutionarily stable strategy adapted to tribal survival in a harsh environment. In some sectors of modern society, contraception and abortion are regarded as infanticide, and science as irrelevant to the question of their legitimacy. But evidence of these practices, even where official opposition is severe, suggests that people in their own minds will legitimize the means available to improve their chances of social, economic, or physical survival.

And lest it be thought that infanticide has no relevance to events in modern America, the conflict between Robert and Peggy Stinson and the medical bureaucracy of a pediatric hospital over the artificial prolongation of the life of a hopelessly handicapped baby reveals the paradoxical twist that the issue may take where scientific medical technology is exploited in the name of unscientific ethical and legal codes.[17] Does the availability of modern biomedical technology, at whatever the cost, make failure to artificially prolong hopelessly impaired and handicapped life homicide? Controversy over the artificial prolongation of comatose life where brain death has occurred has not resolved the issue; and ethics derived from sociobiology are locked into confrontation with medical, religious, and juridical conservatism on the issue of right to death.[18]

The right to life or abortion issue is a forerunner of other biopolitical confrontations yet to come. Public officials are today caught between the conservative position that absolutely opposes abortion as contrary to "natural" law, and the liberal view that public authority has no legitimate control over a mother's "right" to her own body and hence over her freedom to decide for or against abortion. Nature, as perceived through science, tells us little regarding the ethics of abortion; and the circumstance that one-half of the genes of a fetus are contributed by the father, should logically give him a voice in any decision regarding its termination. Yet what if the parents disagree?

An equally sensitive biopolitical issue and one almost certain to become a policy nightmare for democratic government is the challenge to the alleged right of humans to unrestricted procreation. No such right was recognized in many traditional societies. In the United States, the right of society to prevent reproduction of the apparently genetically unfit was once sustained by the Supreme Court.[19] But since 1927, when Justice Oliver Wendell Holmes declared that: "Three generations of imbeciles are enough . . . ," the law has become less certain. Courts have subsequently affirmed a natural and constitutional right to beget children, but have recognized the right of the state, subject to due process of law, to modify this right where science shows the public welfare to be adversely affected.[20] How far this qualified right might be infringed by measures such as those adopted in Singapore to hold down family size is at present conjectural.[21] The Singapore experience may demonstrate the difficulty of sustaining biocratic and ecologic policies as against the pressures of short-range economic advantage. The phenomenal economic growth of Singapore has led to demands for growth of the labor force. The result appears to be a relaxation of population controls. When biological prudence conflicts with economic advantage, economics tends to be the immediate victor.[22]

The United Nations Universal Declaration of Human Rights has been described as " . . . a license for unlimited reproduction."[23] But contemporary opinions that place the procreative rights of individuals above the right of society to protect itself and future generations against the burdens of overpopulation and unfitness are poorly timed. As human populations continue to expand, legal limitations on family size are likely to increase. And as more is learned about human biology, it is at least plausible that a scientific base may be provided for attacking the present inclination of most governments toward benign toleration of indiscriminate reproduction. In any event, law and government are now confronted by new reproductive technologies exemplified by so-called test-tube babies.

This does not mean that, except as verified by biology, there are no longer defensible propositions regarding individual or group behavior. But it may mean that the *obiter dicta* of highly placed persons in politics and in civil and ecclesiastical bureaucracies may be less readily accepted when lacking scientific or empirical support. The practical value of some of the old beliefs and behaviors may be reaffirmed but understood in a new light after biosocial reexamination. The effect of a reorientation regarding human nature is, first, to open to reexamination all propositions not based upon or consistent with objectively tested evidence and, second, to broaden the public agenda relating to human behavior. Both of these effects have radical implications for law and politics and are not without risks or dangers. The first effect would be to change the test of truth from political fiat to an examination of evidence incorporating scientific findings. The second effect, already observable, has been to open the political process to propositions or policies heretofore excluded because of particular moral or cultural inhibitions. In neither case, it should be noted, is amoral or scientistic politics an inevitable consequence. What is implied is a restructuring of social assumptions following a reexamination and redefinition of the basis of policy, substantive and moral.

Who is to do this? I have suggested that it be a public but not governmental task. New types of civic institutions for inquiry, consultation, agenda setting, and consensus building are needed. The concept of "councils of urgent studies" proposed years ago by John R. Platt remains one of the more promising approaches to building an informed consensual base for the policy process. Building on this proposal, I have suggested the development of a new system of civic participation linking the private sector to government that might answer the need for policy innovation.[24]

After Copernicus, advances in the physical sciences revolutionized concepts in popular and official cosmology. Beliefs regarding man's relation to nature were changed and new concepts of celestial mechanics influenced theories of social and political relationships. The visible expression of the new understanding of nature was largely through technology. Physics moved into society and politics overtly through the steam engine, electronics, aerodynamics, and the atom. The mechanistic concepts of physical science also had a powerful, if less obvious, effect upon organization theory and the constitutional structure of governments.

Excepting the conflicts over Darwinism, biology's first inroads upon politics resulted primarily from technological advances, largely biomedical and agricultural. Social implications were less immediately apparent, in part because of the lag time involved in nearly all processes of conceptual change. Biological concepts have not yet led to societal changes comparable in scope and coherence to those resulting from the Copernican revolution, but that is the order of magnitude of the impact that now appears to be impending. Its forefunner appears to be ecology.

Conceptual change of this character almost certainly will not come from the public bureaucracies charged with the protection and administration of the existing socioeconomic order. Changes in public perceptions of the way things are or ought to be are inherently disruptive, especially to guardian institutions. Probabilities appear to be against a consistent adaptation of public bureaucracies to rapidly changing attitudes in politically significant sectors of society toward such matters as motivation, competition, antisocial violence, appropriate education, sex roles, childbearing, and the psychology of work and dependency. How public bureaucracies will accommodate to new concepts of the forces that shape society is far from clear. Notable adaptation has occurred in some judicial courts and political jurisdictions to new biology-based assumptions and behaviors. Attitudes toward the AIDS epidemic and single-parent families are cases in point. Yet opinions differ widely on these and other issues relating to the new biology, and confrontation is more evident than consensus.

Biopolitics and Civic Consensus

In a 1976 bicentennial essay commissioned by the American Society for Public Administration for *Public Administration Review,* I observed that: "The American people are widely committed to the substance of a civic religion."[25] Although not a religion in the conventional sense, there has been an historical consensus that represents what most Americans have believed regarding public morality, the functions of politics, and the proper role of government. The American commitment has been more to stated principles than to their implementation in practice. These principles included the right to self-government, limitation of the powers of government, government by majorities, personal equality, and virtue in public office holders. The practical meaning of these principles was ambiguous in 1776 in relation to a society in which they were widely shared. How much more difficult is agreement on their meaning in a centralized, bureaucratized, technological society whose citizens comprise what Vance Packard has called "a nation of strangers." Will an emerging social biology clarify and confirm the intuitive principles of the old consensus thereby giving them new significance, or will it raise up competing ideologies?

The American consensus now suffers four serious hazards to its continuity. First, it is no longer held in place by an almost universal commitment to the great historical American goal—the settlement and development of the continent. For more than three centuries, the dominant task of America was believed to be fulfillment of God's purpose, revealed to the first man and woman, according to the Book of Genesis, in the words: "Be fruitful, multiply, fill the earth, and subdue it." This manifest destiny was America's great unifying goal, until, once achieved, it left the nation without a common purpose. The search for another unifying vision has led Americans to a mission to make the world safe for democracy and beyond the earth to explore the reaches of outer space. But whatever specific successes these ventures may have had, they have failed to answer their deeper unarticulated need which was to provide a substitute for the frontier and to find a new unifying common goal to which nearly all Americans could subscribe.

The second hazard to the civic consensus is that the social and institutional structure in which it evolved belongs to an irretrievable past. Pro forma, the elements of the American constitutional system would be recognizable by James Madison. In fact, however, the system is greatly changed, responding to a society that differs in almost every respect from the America of 1789. No revitalization movement could restore the old republic, which seems now as irrelevant to contemporary America as is the old frontier. Regardless of rhetorical commitment to the work of the "founding fathers," their misgivings concerning democracy and bureaucracy and their conviction of the need to restrain political power are widely regarded as reactionary when expressed today.

The third hazard to the consensus is lack of empirical support for its premises regarding human nature, derived from the ethical and philosophical doctrines of the seventeenth and eighteenth centuries. Self-evident truths and rights of men, deduced in an age of reason, were based upon intuition, not upon empirical evidence. Intuitive knowledge is not necessarily false, but is most plausible when experience has not compromised its credibility. Today, where the political doctrines of the age of enlightenment are still professed, it is probable that their symbolic value depends upon their remaining unquestioned, their relevance to real life unexamined. "Liberty, equality, fraternity" was a powerful political slogan, but like the American Declaration of Independence, a high degree of rationalizing would be necessary to give it more than rhetorical significance. The old rhetoric still has power, but a new rhetoric with more plausible underpinning, although less inclusive in its appeal, might be a disruptive competitor.

The fourth and most visible hazard is the inordinate expansion in size and power of the federal bureaucracy. Broadly defined, this includes the federal courts and those state agencies controlled in essential respects by federal judges and administrators. Federal courts now develop policy with a purpose to shape public consensus, not to follow it, and to enforce their judgments upon state and local governments in great specificity and detail. The power to withhold funds enables federal agencies to compel local and personal compliance with bureaucratic guidelines and administrative rules.

The reasons for the rise of huge, professionalized, activist, federal bureaucracy is germane to our theme but beyond its immediate scope. Every policy, act, and procedure of the federal bureaucracy reflects the wishes of some Americans. But under the pressure of so large, dominating, and impersonal a government in the urbanized mass society that America has become, it is difficult to believe that the old consensus could ever be realized in practice without a revolution more fundamental than that of 1776.

Yet it would be easy to idealize the old communitarian town-meeting style of self-government and the Jeffersonian concepts of civic virtue.

They were not always the ways in which traditional America actually managed its civic affairs. Moreover, newer forms of citizen activism and methods of enforcing governmental accountability should not be underrated. Nevertheless, the vast size, economic power, and remoteness of decisionmaking authority in both the federal bureaucracy and the corporate structure of business diminish the significance of individuals and their local communities in national affairs. The argument on behalf of big government—that America now has Jeffersonian ends pursued by Hamiltonian means—is wrong in the sense in which it is usually advanced. Jeffersonian ends were largely procedural and decentralized, and Hamiltonian aristocratic republicanism was far removed from the ends or means of bureaucratized equalitarian mass democracy.[26]

The power of the federal bureaucracy has grown in large measure from the inability of America's mass society to agree, as individuals or communities, upon solutions to their civic problems or to resolve them without federal aid. Consensus implies an agreement in principle far deeper than spontaneous responses to weekend crises. An argument could be defended that bureaucracy has arisen as consensus has failed, and that federal mediation and coercion have become necessary because Americans think less alike than formerly. This is partly because those Americans now designated as "minorities" were not fully included in the old consensus. But it is also the disintegration of consensus among the older leadership in America that has opened the way to minority power and legislative factionalism. The weakness of the political parties contrasts with the strength of the bureaucracy. Yet the apparent strength of bureaucracy may be in reality only by contrast to the weakness of civic consensus. Biology would pose a direct challenge to bureaucracy should it become evident, through administrative failings, that human needs and nature were incompatible with centrally directed bureaucratic technostructures. Were large-scale bureaucracy to be seen as a pathological response to social disorder, a search for alternative approaches to the functions of government would be stimulated. At this point, the old civic consensus might gain new relevance (although perhaps differently expressed) if its intuitive conclusions should be found to be consistent with what sociobiologists learned about human nature.

Confrontation with Bureaucracy

Roger D. Masters suggests the following proposition derived from sociobiology: "Large bureaucratic societies are doomed to fail, since our species has evolved specialization for living in face to face groups; participatory democracy is more natural than the centralized state."[27] Analysis of this proposition would be an extended process and would require inputs from more than sociobiology, which is not yet able to help much with prognosis on the future of bureaucracy. From the historical evidence noted in chapter 4 it appears that all large centralized bureaucratic systems have proved ultimately unequal to their missions, and that the life expectancy of those now extant is uncertain.

It does not of course follow that these systems failed *because* they were bureaucratic. Bureaucracy *may* have extended the duration of the empires of Rome and China centuries beyond their otherwise probable point of collapse. And the bureaucratic continuity of the Roman Catholic Church may be explained by its operational decentralization. Nonbureaucratized societies have also failed, although perhaps more often because of external rather than internal factors, for example, by conquest or acculturation by "better organized" societies. Nevertheless, the repeated association of bureaucratic hypertrophy and centralization with social and consensual disintegration suggests cause-effect relationships. But the question arises as to why humanity persists in building big bureaucracies if that form of social structure is unnatural to man and inherently self-destructive.

Masters' proposition has been advanced to illustrate a type of theoretical statement that can be derived from sociobiology. He does not assert the proposition as a principle or law, and concedes that it may be false. There are substantial reasons, however, for believing that present-day bureaucracies, especially in government, are being severely stressed by size, complexity, and by the unprecedented nature of conflicts with which they must cope. But not all public bureaucracies are alike, despite common characteristics. Focus here is upon American public administration which, out of an antibureaucratic past, has developed into a complicated, and in some respects atypical, bureaucratic system. It is far from "monolithic."

Short of confrontation, advances in bioscience have led already to intrabureaucratic conflicts, both latent and overt. Research on smoking has placed public health services in opposition to federal and state departments of agriculture and revenue. Environmental biology has raised adversary relationships between the Environmental Protection Agency and a long list of public (and private) bureaucracies representing economic and technological interests. Studies on the psychology of drug use have differentiated the policy outlook of mental health officials from that of narcotics police. Many more examples could be cited.

But biology may upset both the bureaucracy and the economy if reliable techniques become common to enable parents to select the sex of their children. This possibility, in addition to other refinements in contraceptive technology and a substantial change in traditional attitudes toward children and families, could lead to unforeseen reductions in the national population. Should this occur rapidly and extensively the present structure of bureaucracy and public debt would become an insufferable burden unless anticipating measures were taken to cope with the consequences of demographic change. The advantages of a reduced and stabilized level of population might more than compensate for the difficulties in managing the transition.

Another interesting conjecture would be the consequences of biomedical prolongation of human life and vigor for as much as twenty-five to fifty more years. Implications for employment policy and actuarial science are obvious. Less obvious, but not less portentious, could be the effect upon advances in the social and behavioral sciences. It appears that scientific creativity comes earliest in life in mathematics, physics, and generally in disciplines relying heavily upon analytic methodology. In the social sciences, creativity tends statistically to peak later in life, possibly because accumulating the knowledge and insigts required from synthesis of complex social phenomena takes time.[28] Could it be that one reason for the relative inadequacy of our science of society is that the professional lives of social scientists are too short to master their disciplines? Could it be that twenty-five more years of intellectual vigor added to the lives of creative social and behavioral scientists would revolutionize the status of their disciplines and greatly advance understanding of human behavior? Some people, however, might not regard prolongation of creativity among social scientists as a good thing unless a growth in wisdom occurred simultaneously.

More immediately problematic are conflicts arising between the administrators of statutory policies and sectors of the public who believe those policies to be based upon factual and conceptual error, and therefore not lawfully binding upon self-actualizing individuals. Deliberate publicized violations of law have become common in many advanced countries on grounds of personal freedom of action and a higher morality. As to the mere fact of confrontation, it does not matter whether the weight of scientific evidence is with the nongovernmental public or the bureaucracy. In many cases, scientific facts and their interpretation are in dispute.

The history of the Delaney Amendment of 1958 to the Food, Drug, and Cosmetic Act illustrates the consequences of a mandatory provision inspired by bioscience being written into law in language that scientists generally would not have used. When the amendment was invoked to ban the sugar substitute saccharin, because of carcinogenic properties revealed by laboratory tests, the Food and Drug Administration found itself in confrontation with people who believed that bureaucratic zealotry, in the name of science, offended common sense.[29]

In the past, confrontation has most often occurred when the bureaucracy was (or believed itself to be) guided by science while the public was not. Popular opposition to vaccination, flouridation, mass spraying of pesticides, and banning of laetrile have been familiar cases in point. And findings in ecology, toxicology, epidemiology, and genetics have sometimes reinforced and sometimes undermined federal policies. These biopolitical conflicts have generally been single issue affairs and individually have had no great, or at least obvious, implications for government and the civic order as a whole. Yet advances in social biology could move circumstances beyond this situation, placing public officials in positions of defending not only laws but social and political assumptions which have been losing credibility.

Although there is demonstrable truth in the allegation that bureaucracies tend to be inflexible, unresponsive to novelty, and conservative in relation to political norms, the truth relates more to tendencies than to the behavior of particular bureaucracies. There have been short-term exceptions to general rules. Under policy-focused vigorous direction, bureaucracies can sometimes be swift moving, innovating, and reforming. But bureaucratic organizational and managerial efficiency are widely conceived in mechanistic terms obtained through operations research, statistical control, and the integration of men, machines, and automated processes. Seeking efficiency through mathematical techniques such as cost-effectiveness calculations, management analysts, with some notable exceptions, have tended to discount the biopsychological essentials of human association. As Luther Gulick insists, " . . . any team of people working together for a purpose is an 'organism', not a machine." And he declares that " . . . the test of the organization is not mechanical, as under the laws of thermodynamics; rather it is biological, under the evolutionary concepts of Darwin. The proof is survival."[30]

To organize work on the organic, face to face small group basis that prevailed widely throughout preindustrial society has been viewed as impractical in large systems of administration or production. The assembly line has been a symbol of the efficient organization of work. But mechanistic concepts of organization and management are already under challenge and some manufacturers, notably the Swedish Volvo, have adopted teamwork techniques that appear closer to sociobiological prototypes than to traditional industrial engineering models. The challenge of biology to bureaucracy is not only to the substance of things done, but to the way in which work is organized and administered. What the ultimate effect of changing managerial practices will have on performance is conjectural and may vary among cultures.

Because many sectors of society will be in mutual opposition on many biopolitical issues, the public bureaucracy will face a divided and conflicting constituency. It does so today with respect to laws and policies regarding birth control and abortion, drug use and abuse, genetic research, intelligence tests, pesticides, smoking, health services, responsibilities of physicians, euthanasia, and many more. On these issues legislatures are also divided and face the frustrating circumstance that the conflicts are seldom of a kind in which compromise offers obvious and feasible solutions. The compromises attempted seldom satisfy any of the adversaries, tend to sharpen their distrust of government, and burden public administration with additional regulatory and policing duties that reinforce the image of meddlesome, intruding bureaucrats.

An accumulating effect of these conflicts could be to place the bureaucracy in a "no-win" situation. People, including many "highly educated," tend to be notoriously uninformed, indiscriminate, and imprecise in blaming government for policies that they dislike. Distinctions between governmental programs and agencies are commonly confused, and a generalized animus is directed toward government or bureaucracy. Popular perception of government as incompetent, inefficient, indifferent, and wasteful is widespread. Public confidence in existing institutions and authorities appears to be at a low ebb. The United States seems unable to find satisfactory dispositions (not solutions) for any of the major problems that confront it as a nation. Meanwhile the accelerating growth of bureaucracy and the national debt and the seemingly uncontrollable expansion of bureaucratic authority has begun to arouse antipathy among many citizens who have been sympathetic to the causes for which big government was believed to be necessary. These circumstances would not appear to be favorable to coping with the unprecedented and radical developments inherent in the biological revolution.

If the past is a reliable guide (and in relation to present events it may not be), a confrontation more fundamental than current biopolitical conflicts is likely, although its ultimate expression cannot be foreseen. Western society or, more particularly, American society presents a picture of distrust, disbelief, and widely variant values. There are of course positive countercurrents that lead toward grounds for optimism. Nonetheless there are visible symptoms of disintegration of consensus and growth of contention that historically have often preceded radical change. Out of the concepts, findings, and technologies of the biological sciences could come new formulations of social and ethical theory challenging the fractionated and confused consensus upon which present government and social institutions have been relying. Because the actual form and occasion of such confrontation is unpredictable, common sense and conventional wisdom assure us that it will not happen: in modifications of policy and behavior, perhaps yes; in fundamental social and constitutional change, probably no. Yet historically, revolutions—religious, political, or technological—have never been anticipated by the majority of people, including people most responsible for them. Last to learn has almost invariably been those highly placed in the government in the capital city. De Toqueville's classic lines introducing his history of *On the State of Society in France Before the Revolution of 1789* remain an eloquent comment on social perceptivity:

> Nothing is better fitted to give a lesson in modesty to philosophers and statesmen than the history of the French Revolution; for never were there events more important, longer in ripening, more fully prepared, or less foreseen.[31]

One should not infer that a confrontation between bureaucracy and new biosocial knowledge would resemble historical revolutions. On the contrary, the interlocking character of technology, the control potential of information and communications media, and the relatively high level of popular education are among the factors enhancing the probability that change, however radical, will not follow past scenarios. As with the advent of death and taxes, one may concede the certainty of an event without knowledge of its time, place, or character. The probability of confrontation is therefore best posed as a question, Is it plausible that a century of expanding discovery in the biological sciences will continue indefinitely to have only a marginal effect upon theories and practices of society and government? Will cumulating scientific insight into human personality and social behavior eventually force rejection, redefinition, or reaffirmation of the principles that underlie the structure and functions of government in America and the Western world?

In a retrospective on his *Psychopathology and Politics,* Harold D. Lasswell wrote that:

> We are in the early stages of an explosive development in every field of scientific knowledge, including knowledge of men's changing value pre-dispositions and gratifications through time. The tempo has risen markedly, though not as rapidly in the knowledge of mastery of self as in the command of nature. As the discrepancy grows, our tenure of life as a species grows more precarious.[32]

It is widely understood that the sciences of human behavior have lagged behind the sciences of the material world but the implications of this imbalance are far less widely appreciated. It is also widely believed that advances in our understanding of man and society could correct this dangerous imbalance in knowledge, and enable humanity to become a safer custodian of its destiny. But it has been somehow assumed that the growth of knowledge could only have positive, ameliorative effects. That true knowledge of man's inner nature might exact heavy psychological and ethical costs has not generally been foreseen. And it is possible that the additional knowledge required for self-mastery of man and society may be even more costly to apply than to acquire.[33]

True knowledge of self and society will almost surely be dangerous, although how much more dangerous than ignorance or error is uncertain.[34] More certain is the prospect that societies will be divided upon questions of whether to retain and advance present levels of technoscientific culture by paying the social costs required, whether to fall back to a simpler way of life that can be administered through minor modifications of traditional methods, or whether to develop new social arrangements consistent with a more profound understanding of human nature. A potentially advantageous result of advances in biobehavioral science could be greater, and much needed, understanding of the psychobiological characteristics of individual humans that would affect their suitability as political leaders. This is a frontier area of biopolitical research and, should it lead to reliable information, a major social problem could arise over efforts to apply it to candidates for public office.[35]

These may not be happy choices, but they may be the alternatives. Risks are unavoidable. But the society that succeeds in meshing an advancing knowledge of self and society to an ability to manipulate the material world within the limits laid down by nature will achieve the best prospect of survival. At what level of organization and technology this meshing would most likely begin is not predictable today.[36] Phaseback or reorientation may be necessary if nature firmly limits the tolerance of all men for organizational size and complexity. And if individuals and peoples should be found ultimately to differ greatly, unchangeable, and hence unequally in this ability, then a social future very different from the present may be in prospect. The limiting factor in humanity's future may be humanity itself.

Notes

1. For more detailed accounts of the impact of biology on society, see notes on chapters 1 and 2 and the following: Richard L. Means, "Sociology, Biology, and the Analysis of Social Problems," *Social Problems* 15 (Fall 1967): 200–212; G. R. Taylor, *The Biological Time Bomb* (London: Thames and Hudson, 1968); Paul R. Ehrlich, "The Biological Revolution," *Center Magazine* 2 (1969): 28–31; F. J. Ebling, ed., *Biology and Ethics* (London: Academic Press, 1969); and Watson Fuller, ed., *The Social Impact of Modern Biology* (London: Routledge and Kegan Paul, 1970); and Elliott White and Joseph Losco, eds., *Biology and Bureaucracy:*

Public Administration and Public Policy from the Perspective of Evolutionary, Genetic and Neurobiological Theory (Lanham, Maryland: University Press of America, 1986).

2. See *New York Times*, 29 January 1986. The tragedy of the Challenger made headlines in nearly all newspapers.

3. There is a wide range of reactions to sociobiology—ethical, scientific, and polemical. No attempt here is made to provide a bibliography of the literature. Note, however, the following: Edward O. Wilson, *Sociobiology: The New Synthesis* (Cambridge: Harvard University Press, 1975); and Charles L. Lumsden and Edward O. Wilson, *Genes, Mind, and Culture* (Cambridge: Harvard University Press, 1981). Sociobiology in a generic sense was current at least a decade before Wilson's publication. See A. M. Guhl, "Sociobiology and Man," *Bulletin of the Atomic Scientists* 21 (October 1965): 22–24. Distinction between sociobiology in the generic sense and its special connotation is emphasized by Gerald Holton in "Sociobiology: The New Synthesis?" *Newsletter on Science, Technology and Human Values* 21 (October 1977): 28–43. See also George W. Barlow and James Silverberg, *Sociobiology: Beyond Nature/Nurture, Reports, Definitions, and Debate*, AAAS Selected Symposium (Boulder, Colorado: Westview Press, 1980); Scott Gordon, *A Critique of Sociobiology*, discussion no. 346 (Kingston, Ontario, Canada: Queen's University, Institute for Economic Research, undated); Albert Rosenfeld, "Sociobiology Stirs a Controversy over Limits of Science," *Smithsonian* 11 (September 1980): 73–80, Arthur L. Caplan, *The Sociobiology Debate: Readings on Ethical and Scientific Issues* (New York: Harper & Row, 1978); Bernard D. Davis, "The Importance of Human Individuality for Sociobiology," *Perspectives in Biology and Medicine* 26 (Autumn 1982): 1–17; and a response by E. O. Wilson, "Sociobiology, Individuality, and Ethics," *Perspectives in Biology and Medicine* 26 (Autumn 1982): 19–29. Closely relevant is the June 1984 issue of *Zygon* (vol. 12, no. 19) carrying a series of articles on "The Challenge of Sociobiology to Ethics and Theology."

4. E. O. Wilson, *On Human Nature* (Cambridge: Harvard University Press, 1978), 10; and a review by Robert Nisbet in *The American Spectator* (April 1979): 30–31. See also, chapter 27, "Man: From Sociobiology to Sociology," in Wilson's *Sociobiology: The New Synthesis* (Cambridge: Harvard University Press, 1975).

5. Wilson, *On Human Nature*, 13.

6. In David P. Barash, *Sociology and Behavior* (New York: American Elsevier, 1977), 13.

7. Roger D. Masters, "The Political Implications of Sociobiology," Part I, Inclusive Fitness Theory in the Western Tradition (Paper presented at the second annual meeting of the International Society of Political Psychology, held in Washington, D.C., 24–26 May 1974, Mimeographed), 6.

8. E.g., "A Little Matter of Birth Rates," in Samuel Lubell's *The Future of American Politics* (New York: Harper and Bros., 1951), 28–34. The principle that less may lead to more may be illustrated by juncture of low birth rate and high income and occupational status among identifiable American ethnic groups. [Note implications of statistical comparisons of average age and status of identified ethnic groups by Thomas Sowell in "Myths about Minorities," *Commentary* 68

(August 1969): 33–37; and his more comprehensive *Essays and Data on American Ethnic Groups* (Washington, D.C.: Urban Institute, 1978).] But for international competition, see the conclusions of Katherine Organski and A. F. K. Organski in *Population and World Power* (New York: Alfred A. Knopf, 1961).

9. Masters, "The Political Implications of Sociobiology," Part I, 7.

10. Cf. Garrett Hardin and John Baden, eds., *Managing the Commons* (San Francisco: W. H. Freeman, 1977).

11. Barash, *Sociology and Behavior,* 81.

12. Cf. Milovan Djilas, *The New Class: An Analysis of the Communist System* (New York: Praeger, 1957); and Hedrick Smith, *The Russians* (New York: Quadrangle Books, The New York Times Company, 1976).

13. John Maynard Smith, "The Evolution of Behavior," *Scientific American* 239 (September 1978): 184–85. The author explains: "The strategies I was seeking had little to do with the optimum strategies with which traditional game theory deals . . . . I hoped to determine an evolutionarily stable strategy: a strategy that if most of the members of a large population adopt it, then no mutant strategy can invade the population. In other words a strategy is evolutionarily stable if there is not mutant strategy that gives higher Darwinian fitness to the individual adopting it." See also "Evolution and the Theory of Games," in T. H. Clutton and Paul H. Harvey, *Readings in Sociobiology* (San Francisco: W. H. Freeman, 1978), 258–70.

14. Luther Gulick, "Democracy and Administration Face the Future," *Public Administration Review* 37 (November–December 1977): 709.

15. Mildred Dyckemann, "The Ecology of Hypergynous Mating Systems," in *Social Science Information* (Paris: International Social Science Council, 1979).

16. Betty Meggers, *Amazonia: Man and Culture in a Counterfeit Paradise* (Chicago: Aldine-Atherton, 1971).

17. "On the Death of a Baby," *Atlantic Monthly* 244 (July 1979): 64–72.

18. Earleen H. Cook, *Euthanasia and the Right To Die: The Medical and Legal Viewpoint* (Monticello, Illinois; Vance Bibliographies, 1982); Madeline M. Nevins, ed., *A Bioethical Perspective on Death and Dying: Summaries of the Literature* (Rockville, Maryland: Information Planning Associates, 1977); Joanne Lynn, ed., *By No Extraordinary Means: The Choice To Forgo Life Sustaining Food and Water* (Bloomington: Indiana University Press, 1986); Gerald A. Larue, *Euthanasia and Religion: A Survey of the Attitudes of World Religions on the Right-To-Die* (Los Angeles, California: The Hemlock Society, 1985); and A. Carmi, et al., eds., *Euthanasia,* includes bibliographies (New York: Springer-Verlag, 1984).

19. Buck v. Bell, 274 U.S. 200, 47 S.Ct. 584 (1927).

20. Cf. Skinner v. Oklahoma, 316 U.S. 535 (1942); and In Re Cavitt 157 N.W. 2d. 171 (1968).

21. Wan Fook Kee and Margaret Loh, "Singapore," *Studies in Family Planning* 5 (May 1974): 163–65.

22. Personal communication from the Hon. Devan Nair, former president of Singapore.

23. A. S. Parks, "The Right to Reproduce in an Overcrowded World," in *Biology and Ethics,* ed. F. J. Ebling (London: Academic Press, 1969), 109.

24. Lynton Keith Caldwell, *The President as Convenor of Interests: Policy Development for the Twenty-First Century,* Distinguished Public Affairs Research Lecture (Bloomington: School of Public and Environmental Affairs, Indiana University, 18 October 1984).

25. Lynton K. Caldwell, "Novo Ordo Seclorum: The Heritage of American Public Administration," *Public Administration Review* 36 (September–October 1976), 481.

26. However, J. Donald Kingsley states the issue correctly in his book review essay, "Political Ends and Administrative Means: The Administrative Principles of Hamilton and Jefferson," *Public Administration Review* 5 (Winter 1945): 87–89. Kingsley emphasized the importance of public consensus in the effective implementation of policy.

27. Roger D. Masters, "The Political Implications of Sociobiology," Part III, Birth Control, Celibacy, and Inclusive Fitness. Distinction should be made between the survival of bureaucratic systems and of individual bureaucratic agencies. Herbert Kaufman has examined the almost unexplored prospects of the latter in *Are Government Agencies Immortal?* (Washington, D.C.: Brookings Institution, 1976). How much light sociobiological analysis throws on the behavior of either level of bureaucracy is conjectural, especially given the gross deficiency in the kinds of information needed. It is conceivable that sociobiological analysis may help to explain large-scale social trends but have little applicability to the agency life histories which Kaufman examined.

28. Evidence for age differential in creativity in the sciences is based particularly on work by Harvey C. Lehman. See *Age and Achievement* (Princeton, New Jersey: Princeton University Press, 1953), and "The Age Decrement in Outstanding Scientific Creativity," *American Phychologist* 15 (February 1960): 128–34. See also Donald C. Pelz and Frank M. Andrews, *Scientists in Organizations: Productive Climates for Research and Development* (New York: John Wiley, 1966), chapter 10.

29. Cf. W. A. Tomasson, "Dangerous to Your Health: Saccharin, Cancer, and the Delaney Clause," *Atlantic Monthly* 243 (June 1979): 25–26, and comment in the September issue, 31–32.

30. Gulick, "Democracy and Administration," 709.

31. Alexis de Toqueville, *On the State of Society in France before the Revolution of 1789,* trans. Henry Reeve (London: John Murray, 1856), 1.

32. Harold D. Lasswell, *Psychopathology and Politics* (New York: Viking Press, 1960), 319.

33. Cf. Wilson's concluding observation in *Sociobiology: The New Synthesis,* 575: "To maintain the species indefinitely we are compelled to drive toward total knowledge, right down to the levels of the neuron and gene. When we have progressed enough to explain ourselves in these mechanistic terms, and the social sciences come to full flower, the result might be hard to accept."

34. Cf. Van Rensselaer Potter, chapter 5, "Dangerous Knowledge: The Dilemma of Modern Science," in Potter's *Bioethics: Bridge to the Future* (Englewood Cliffs, New Jersey: Prentice-Hall, 1971), 69–82.

35. Cf. Lasswell, *Psychopathology and Politics,* and Meredith W. Watts, ed., *Biopolitics: Ethological and Physiological Approaches, New Directions for Methodology*

of Social and Behavioral Science (Seves, New York: Jossey-Bass, 1981). Behavioral genetics seems certain to attract continuing investigation even though it arouses apprehension regarding possible implications of its findings. Note Lawrence Taylor, *Born to Crime: The Genetic Causes of Criminal Behavior* (Westport, Connecticut: Greenwood Press, 1984).

36. For an international perspective of the ethical challenges that the new biology poses for society and government, see the report by Bruno Ribes, *Biology and Ethics: Reflections Inspired by a UNESCO Symposium* (Paris: UNESCO, 1975), 178. The symposium, "Meeting of Experts on the Ethical Problems Posed by Recent Progress in Biology," held at Varna, Bulgaria, 24–27 June 1975, was attended by "biologists and moralists." It may be indicative of widespread underestimation of the political significance of the issues, that public officials do not appear to have been present. Note also the previously cited work edited by F. S. Ebling.

8
Biotechnology: Options, Benefits, and Risks

Biotechnology may be defined with a general or a specific reference, although the latter tends to be preferred in scientific literature. This specific or specialized meaning refers to the use of living materials as tools not only to manipulate the forms or functions of other living materials, as in recombinant DNA techniques, but also as agents for control of water pollution, reduction of frost damage to crops, and the recovery of metallic ores. The definitive condition of this special meaning is the instrumental use of living material.[1]

The more broadly inclusive use concerns the manipulation of the behavior of living material by whatever technology proves effective. Such techniques may employ biological, chemical, or mechanical agents, frequently in combination, but living materials (usually organisms) do the "work." These biology-driving techniques play important roles in agriculture and medicine. It is thus apparent that the boundaries of biotechnology in the broad sense cannot clearly be demarked. It is less evident, but no less true, that the definition of biotechnology in the narrow or specialized sense is not unequivocal. For example, if a microorganism is used to synthesize a chemical compound which is subsequently applied to altering some life form or process that, presumably, is biotechnology. Were the same compound to be synthesized chemically in a laboratory and applied to living material, would that then be biotechnology or chemical technology? When laboratory animals are used to manufacture vaccines or otherwise employed as instruments in biological research, does their use constitute biotechnology? Few biological techniques, moreover, can be applied without the use of mechanical or electronic devices.

There is little point to laboring definitions except to show that the range of knowledge and technique that may be applied to changing the forms or functions of living material cannot be defined with precision. The substance of this book does not require that biotechnology be precisely defined, and both its general and specialized meanings will be used as appropriate to each case in point.

Some forms of biotechnology have been around for millenia. Long before the causal mechanisms were fully understood, humans used microorganisms to make beer, wine, cheese, and bread. People crossbred animals and plants to obtain preferred characteristics and only in our time have they used high resolution microscopy and other advanced technical substitutes for trial and error.[2] Biotechnology has become an increasingly volatile factor in society since it has been based on science—conceptual and methodological. Molecular biology and microbiology account for the more spectacular advances in biotechnology, but physiology and ecology have also contributed knowledge and technique that have extended human capabilities to influence natural processes.

The growth of science-based biotechnology has required simultaneous advances in chemistry, physics, instrumentation, and engineering. Biotechnology is carried on with the assistance of other science-based technologies (e.g., chromatography, microscopy, computer technology) and, of course, within a society from which its purposes are derived and within which its impacts are experienced. Our concern here is less with the scientific or operational aspects of biotechnology, and more with its influence on public policies. It is especially the growth of biotechnology in the larger sense that has given pertinence and importance to the concept of biocracy in present-day society.

Opportunities and Limitations

Biotechnology, however defined, provides a growing range of options and has made possible new science-based industries. It may play an important role in the production of food, in the conservation of natural resources, in the control of pollution, and in the recycling of materials. Its most widely advertised opportunities relate to human health and sexual reproduction, but it also offers promise of significant contributions to food production and energy.

Opportunities opened by biotechnology provide pathways to benefits otherwise largely unattainable. Among these benefits must be counted opportunities for the advancement of human understanding of life, its processes, and relationships. Biotechnology, in effect, tests concepts and methods derived from basic research. It may also contribute to basic research through the invention and perfection of research techniques. For example, gene splicing and recombining DNA are processes in basic biological research, but require technical implementation.

The list of opportunities opened by advances in biotechnology is extensive and growing. In medicine, among other benefits, is the development of cheaper and more effective vaccines and antibiotics, tracers for the diagnosis of disease, and techniques for curing genetic disorders. In agriculture, biotechnology is developing crops, which make their own fertilizers and pesticides and that are tolerant of dry or saline soils. Biotechnology also assists the production of fuels such as methane, hydrogen, and fuel alcohol which are indefinitely renewable and less polluting than hydrocarbons. It also assists the extraction of underground oil residues from wells no longer productive by conventional methods. Biotechnology is playing an increasing role in developing raw materials for the manufacture of food stocks for domestic animals, and the fabrication of artificial fibers, adhesives, paints, and plastics. Microbial methods have been developed to extract metals from rock. Domestic and industrial wastes have been treated by microbial techniques which are now being utilized in the cleanup of oil spills, among some other forms of chemical pollution in the environment.

Substances manufactured through biotechniques tend to be biodegradable to an extent the materials developed from metals and hydrocarbons (e.g., aluminum cans and styrofoam cups) are not. Bioengineered cultivated plants, with greatly reduced need for fertilizers, water, herbicides, and insecticides, obviously have great environmental advantages. Biogenerated fuels have the advantage of renewability and fewer hazards to health, safety, and environmental quality than oil, coal, and natural gas. Viewed in long-term perspective and with regard to the total costs of fuel cycles, bioenergy may hold great economic advantages.

Long-term advantage unfortunately does not imply short-term practicality. The industry-economy infrastructure of modern society has evolved around the exploitation of fossil fuels. There are weak incentives and strong disincentives to early development of bioenergy technologies. Other areas of potential application hold earlier promise. Biotechnology, nevertheless, may provide opportunities to escape economic entrapment in practices and processes that have become seriously problematic.

American agriculture produces more food than the free market demands, with resulting decrease in farm prices and increase in government price support payments. Through biotechnology, agriculture could be shifted in the direction of plant production for industrial manufacturing and energy. The American economy as a whole could benefit. Diversification in crop production should have beneficial effects on international relations, as for example through easing competition among the common market countries of Europe. Biotechnical innovation in agriculture could also affect economic relationships between industrially advanced countries and those that depend heavily upon export of raw materials and especially agricultural products. The question of who produces what and in what amounts would seem to require more international cooperation than it receives at present. Diversification of opportunities through biotechnology could facilitate this development and lessen the risk of international antagonisms.

Enlarged opportunities for economic improvement are thus inducements to invest in research and development in biotechnology. Where biotechnology leads to improved health and environmental quality, a further advantage may result for the society that is able to obtain beneficial returns from investments in life science research and development. But not all opportunities necessarily lead to benefits, or at least not benefits for all who may be affected. As will be noted in chapter 9 there are liabilities as well as benefits in connection with international commerce in agricultural products of biotechnology. Opportunities may also lead to harm and hazard as in biological warfare. Inadvertent harm may result from well-intentioned innovation as in the use of growth hormones in domestic animals.

The achievements of biotechnology have encouraged the belief that they may be indefinitely expansible. In addition to its contribution to health, environmental quality, and economic welfare, biotechnology has also contributed substantially to science fiction.[3] Yet because we do not know the limits of our biotechnical capabilities it does not follow that everything is possible. The cloning of humans and the engineering of genius are not likely to happen in the foreseeable future.

Limitations on biotechnology are both natural and cultural; the former are limits inherent in nature. Life, for all its range and diversity, cannot be manipulated or transformed without limit. Humans may stretch the parameters laid down in the course of evolution; they have not yet changed the basic conditions under which life exists. In 1913 a biochemist, Lawrence J. Henderson, published a treatise entitled *The Fitness of the Environment: An Inquiry into the Biological Significance of the Properties of Matter* in which he reported his investigation as to whether life could exist under conditions other than those prevailing upon the earth.[4] He concluded that the physical-chemical conditions that have prevailed on earth since life began are those necessary to sustain it, and that other physicochemical regimes would not be life supportive.

Environmental changes have occurred throughout geologic time, but they have always been a kind to which some forms of life could accommodate. Microorganisms have been found that endure hot spring heat or polar cold, but the so-called higher life forms appear to be bound by more narrow evolutionary limits. We have discovered that life exists in the total absence of sunlight and under enormous pressures in the vicinity of volcanic fissures in the deep seabed. Some of these life forms are nourished directly by chemicals in the volcanic gases without photosynthetic processes. We have not ascertained for certain the outer limits of life and hence of the options of biotechnology. We have, however, discovered the natural limits for some life processes and are thus able to be wiser in our efforts to manipulate nature.

Cultural limitations are more apparent, but also more flexible restraints upon the possible. They are socioeconomic, psychological, religious, legal, ethical, and aesthetic in various combinations. Not only do they limit what people are willing to accept from advances in the life sciences, but they also limit perceptions of what may be possible. It is a commonplace of human experience that people do not always take advantage of the opportunities available to them. For example, it has been alleged that the plant world contains sources of potential foods, fibers, and medicinal products that have never been explored or exploited.

Benefits of biotechnical innovation have been handicapped by commercial and social inertia. For example, newly developed foods or fibers must win consumer acceptance before they become profitable products in commerce. Economically feasible means of distribution must be available before they can obtain consumer acceptance. Socioeconomic factors obviously influence the implementation and spread of biotechnologies and their products. A favorable bias in a society toward novelty and a market structure that allows for innovation provide favorable conditions for advancements in biotechnology.

Centrally directed economies appear to be less responsive to biotechnical development, in part because of concern to protect their investments in status quo technology. Technological innovation may challenge conventional products and production methods and may therefore have disruptive effects upon some parts of an established economy. Few governments or large corporate enterprises are likely to develop enthusiasm for developments that diminish the value of their investments in products and processes that biotechnology might make obsolete when the profitability of biotechnical innovations is uncertain. In the long run, however, cultural inhibitions regarding biotechnology are likely to be overcome where personal preferences are served and where persons perceive a selective advantage in an innovation regardless of official disapproval. The spread of contraceptive techniques over objections of church and state is a documented case in point.

Costs and Risks

Benefits are seldom, perhaps never, wholly free. The benefits of biotechnology cannot be obtained without the payment of some costs, but all costs may not be apparent at the time that benefits are obtained. Some costs may be unforeseen or delayed and in some cases the benefits may in retrospect appear to have not been worth the ultimate price.

That, however, is a risk often taken in the course of technological innovation.

Costs are incurred to provide the conditions for biotechnological innovation; other costs may follow as consequences. To obtain the benefits of biotechnology, investments must be made in research and development, in education and training, and often in follow-up monitoring and control. There must be provisions for laboratories, equipment, and test facilities. Scientists and technicians must be prepared to work with varieties of chemicals and living material including sometimes dangerous substances and organisms. There must also be public education for the reception and wise use of biotechnical innovations. Legal procedures in patenting, licensing, permitting, verification of health and safety requirements, and possibly defending against lawsuits add costs.

Risks, therefore, are part of the costs involved in biotechnological innovation and may be both physical and monetary. For example, we have noted that bioengineered products may disrupt existing consumer and market relationships. There may be a probability that society as a whole would gain, but a corresponding possibility that some people will lose as the result of new developments. For any new process or product there is the risk of unforeseen consequences and these may or may not be desirable. To the extent that unwanted consequences may be foreseen or suspected, there are costs of prevention or containment of anticipated effects. Payment of these costs may be preferable to foregoing the anticipated benefits; some people may be prepared to assume the risks to obtain the benefits, but others may claim compensation for alleged damages or ask the courts to enjoin biotechnical processes entailing unacceptable risks.[5]

Social problems may arise, moreover, when the benefits are enjoyed by one sector of society and the costs borne by another. There are known physiological risks associated with some aspects of biotechnology that ought somehow to be accounted for in balancing total benefits and costs. People have experienced fears of unforeseeable and conjectural dangers in connection with biotechnology, some of which may be attributed to poor information and exaggerated claims and alarms by misinformed journalists, or occasionally by misconstrued warnings by scientists. Part of the fear of biotechnical innovations, particularly in the area of microorganism techniques, may be attributed to the probabilistic nature of science which seldom affirms that the risk associated with an event in nature can be reduced to point zero. Fail-safe systems and procedures seem to be vulnerable to human failure, and techniques in which danger is present cannot be so contained as to be unequivocally danger-free. Thus public insistence on 100 percent freedom from risk in biotechnical innovations is to demand what cannot be guaranteed and is widely disregarded in other aspects of technology (e.g., automobiles and air-conditioning).

It may be argued that except in cases of high potential danger the costs of biotechnology gone wrong are not necessarily greater than the opportunity costs of biotechnologies not developed. If biotechnology could yield the greater part of the benefits that now appear to be possible, would it not be economic shortsightedness to fail to make the necessary investments to obtain them? For example, if biotechnology, through more effective and efficient methods, could greatly reduce costs in agriculture and pollution control it would be wasteful to persist in present practices if biotechnical innovations could replace them. Even now we can foresee future needs in energy and food production that we have no good prospect for meeting under present circumstances. As we learn more about the relationships between nutrition and health, the effects of the things that we eat become more significant not only for enjoyment of life itself, but as a means of reducing the social and economic costs of illness and premature death. In addition, changes in the quality and character of the earth's atmosphere owing to the emission of carbon dioxide, sulfates, and trace gases from industrial activities may adversely affect future climates and prospects for life on earth. Biotechnology might be developed to prevent or offset these trends which at present are already measurable.

Equities and Economics

Far more common in the politics of biotechnology are controversies over the allocation of burden and benefits resulting from biotechnological innovations and their intrusion upon moral conduct. Because biotechnical advances create new opportunities, risks, and relationships, issues in equities arise. Biotechnical innovations are evaluated very differently by various elements in society. For example, to women, parents, and would-be parents, techniques such as in vitro fertilizations, artificial insemination, amniocentesis, and abortion are likely to have a significance not shared with other sectors of the population. Some people oppose the availability of these technologies on grounds of ethics and morality, whereas others protest denial on grounds of equity. Religious organizations have condemned them on grounds of doctrine and urged governments to prohibit them. For these forms of biotechnology, exercise of free choice is placed on the defensive. Market forces do not allocate for ethics or equity, and government is therefore frequently drawn into controversies regarding issues of morality, freedom, and control in the use of biotechnologies. Government of course does not operate in an environment of neutrality and so political and economic forces are also involved in the shaping of public decisions.

The course of biotechnical innovation is complicated by differing interests prevailing at different levels in society, many of these interests involving considerations of equity. For example, on issues involving policies regarding human genetics there are questions of equitable treatment between individuals and among various groups: occupational, marital, sexual, physiological, ethnic, and religious. There are, of course, many complex interrelationships among these categorical groupings. It is not always the nature or probability of risk or inequity that is at issue among these groups; as often it is the comparative incidence or relativity of risk or inequity. There are of course differences of both fact and opinion over the acceptability of risks and whether the benefits outweigh the hazards. Genetic screening and manipulation and biological weaponry tend to be of this character and have high level of emotive response.[6]

Biotechnology issues may be involved in superficially unrelated issues such as the sitings of plants for sewage disposal or for the destruction of toxic materials, activities which almost everybody agrees are necessary and almost nobody wants nearby. The siting issue illustrates controversy over the allocation of risks where there is general agreement that the benefits are generally shared. For example, localized groups have objected to being put at risk by crop dusting with biocides to protect agricultural products, the benefits of which will be enjoyed by owners and consumers who may not live in the area directly affected. Concern for distribution of risks is therefore frequently linked to controversy over net allocation of benefits. Major concern in the recombinant DNA dispute was that the benefits of experiment would accrue to a small group of research biologists, whereas the entire public might be put at risk if harmful organisms escaped laboratory containment.[7]

Another area of controversy over equities concerns the respective responsibilities of universities, industries, and government in biotechnical research and product development and marketing. Issues of freedom of research, patent rights, science versus trade secrecy, are among potential subjects of biopolitical controversy. Rights are also involved in access to vaccines of limited supply, to organ transplants, to artificial organs in heart and kidney disease. Equities cannot be assigned in the absence of agreed premises, and for many bioethical issues (e.g., fetal life and euthanasia) incommensurable premises are common. The circumstance complicates the almost inevitable necessity for rationing medical treatment.[8] Behavioral alteration through genetic or biochemical technologies are widely condemned as invasions of human rights; commitment to psychiatric hospitals or imprisonment being preferred alternatives.

As developed at greater length in chapter 9, issues of equity have arisen between nations, notably between those advanced in biotechnology and those with minimal capabilities. But there are also issues in equity in the way in which the relationships between governments and biotechnical innovations are assessed in respective countries. For example, Japan has made rapid and far-reaching innovations in many phases of biotechnology, exceeding some advances made in the United States.[9] The Japanese government is a major patron of biotechnology, and working relationships between the government of Japan and the private research, development, and manufacturing sectors are close. In contrast, in the United States biotechnological innovation is generally regarded as a responsibility of the private sector with minimal government involvement beyond financial support and protective regulations.

In fact, government involvement in biotechnology has been extensive from its earliest beginnings in agriculture and public health. The large differences between governmental and private relationships in Japan and in the United States may be more apparent than real; more matters of style than substance. It has been argued, however, that reliance wholly upon private initiative for beneficial products from biotechnology is inequitable because initiative is limited by the necessity of obtaining a profitable return, and demand, defined in the economic sense, relates to the ability to pay, not to need. It follows therefore that biotechnological innovation will primarily serve the economic interests of the technologically advanced nations. Economic development alone, it is argued, can never begin to meet the needs for all of the people especially in technologically backward countries of the so-called Third World. To address these issues extensive efforts described in chapter 9, are being undertaken by international organizations.

In a world in which there are great differences between the conditions desired and needs of people, it is difficult to know how equities should be understood. Some costly applications of biotechnology are restricted to a certain relatively small number of individuals. Investments through government, inevitably influenced by pressure politics, may result in expenditures greatly disproportionate to any rational calculus of public benefit.

Most of the equities issues in biotechnology that do not involve physical risk involve economics. A major consideration of equities in economics is that market forces are limited to (hoped for) profitable transactions. To the extent that biotechnology (with proper safeguards) can reduce the costs of goods and services and assist their wider distribution, equities correspondingly will be served. Industrial biotechnology thus far appears to have no pronounced effect upon the aggregate number of jobs in society. It may, however, have some effect in the displacement of individual employment as between industries, for example, from the chemical industry or universities to biotechnology firms. If through research in plant genetics there would be propagation of plants with greatly reduced needs for fertilizer, herbicides, and insecticides, industries manufacturing these agricultural products might be significantly affected.

Net economic benefits to the locales of biotechnological industries are not easy to estimate realistically. It is characteristic of these industries that their location is not limited by the availability of natural resources and proximity to market that influenced the location of heavy industries of the late nineteenth and twentieth centuries. Many forms of biotechnology can be carried on where its managers and scientists prefer to live. A collateral consideration is the presence of other and related high technology industries and especially of large research universities. Beyond that, the natural and cultural environment are factors, and it is hardly surprising that those mobile industries tend to be concentrated in California near the coast and in the vicinity of Massachusetts Bay in New England. It is not to say that biotechnology will not flourish in midcontinent America, but efforts in that region may need to take account of its widely perceived if often overstated cultural and environmental deficiencies.

Fractionated Policy Constituencies

The new biotechnology has developed and expanded far more rapidly than has popular awareness of its implications, or of institutional provisions to monitor or where need be to control it without retarding its beneficial development. The diversity of biotechnical applications precludes a general policy for either promotion or control. There is a corresponding diversity of clients for biotechnical innovation ranging from agriculturalists to zookeepers. Even in areas of particular application as, for example, in human genetics and medicine, the parties that are interested and concerned are numerous and often in disagreement.

At least six large and diverse groups may be identified. These are the parties whose special interests are expressed in efforts to shape public policies. The general public is affected but, except on emotion-arousing issues, tends to be the passive recipient of biotechnical innovation and policy decisions. Moreover, popular understanding of the issues is handicapped both by lack of information and by misinformation. Popular journalistic treatment of life science issues tends with some exceptions, to accentuate sensational and controversial issues, and is especially deficient in perspective and in depth of analysis. Informed and thoughtful journalism can be found, but is largely restricted to such publications as *Perspectives in Biology and Medicine, Politics and the Life Sciences,* and those of the Hastings Center, Institute of Society, Ethics, and the Life Sciences, that do not circulate widely among the general public.

First among the several parties of interest in public policy for the life sciences are the scientists themselves. But there are many diversities among them—philosophical as well as disciplinary. Molecular biologists and ecologists have been known to differ widely on certain biotechnical applications. Biochemists and ornithologists may have very different viewpoints on uses of life science knowledge. Individual scientists in the same field of work may differ strongly on issues of morality and policy; differences especially marked in relation to both research and application in biomedicine and human genetics. Biologists in agriculture often promote policies that are regarded as short-sighted and destructive by some soil scientists, botanists, and zoologists. Scientists are also citizens and some are committed to social and political ideologies that influence their views on biotechnical applications and even on the propriety or legitimacy of particular lines of research. The Boston-based Committee for Responsible Genetics has taken strong left of center positions on biotechnological and sociobiological issues. Opinions differ sharply on research into the genetics of intelligence, on indentification and restriction of carriers of the hepatitis B or AIDS viruses, and on the right of terminally ill patients to refuse life extending treatment. There are of course many other areas of disagreement.

A second group directly concerned is the industrial sector which divides into several subsets, notably biomedical and agricultural. Manufacturers of biotechnical products, drugs, medicines, and bionic devices have an obvious stake in public policies and regulations. The new biogenetic firms are merely the latest arrivals in the industrial sector. There are in addition a large number of firms, including advertisers, whose interests are affected by public policies for the life sciences; the cosmetic and tobacco industries including their advertisers being an obvious cases in point. An Association of Biotechnology Companies has been formed to advance common interests and to protect its members against interruptions of research activities.

A third group of concern relates to the ethics of biology, to biological research subjects, methods, and applications. The antivivesectionists, opposing the use of animals in experimental biology, are among the oldest of the concerned groups. In early years of research in anatomy, physiology, and medicine, a large sector of the public disapproved of close examination of the human body. Even bodies of dead humans, needed for investigation and instruction were obtainable, only by stealth and contrary to the prevailing legal and religious prohibitions. Today there is organized public concern with bioethics and numbers of institutions whose business it is to observe, report, and comment on ethical and public interest aspects of developments relating to the life sciences. Religious bodies are, of course, among these.

An extension of this category constitutes a fourth group of ethical concern and includes those organizations whose interests are joined to considerations of human rights and law. Prominent among these are civil rights advocates and groups asserting the right of individual privacy over social concerns. Organizations in this category may both defend and oppose biotechnology depending upon its impact on the alleged rights of individuals and particular groups. The legal profession has a variety of concerns ranging from medical malpractice suits and civil rights violations to proprietary and patent rights for biologically engineered products. When biotechnical issues are taken into the judicial courts, lawyers may be found on all sides of the controversies.

A fifth group includes the diverse health-related professions, physicians, hospital administrators, biomedical researchers and the faculty members of schools of medicine and public health. Some of the concerns represented in this category are analogous to those of the industrial sector, notably those of health care maintenance organizations, hospitals, clinical and research laboratories, and some sectors of the food and cosmetic industries which, although not directly engaged in biotechnology, are recipients and purveyors of health-related products developed through biotechnology.

A sixth, and not necessarily a final category of interest, is that of environmental concern. Here, as in the foregoing groups, one finds diverse and contrasting viewpoints. Some persons regard the natural environment as properly subject to manipulation in the human interest and would use biotechniques to ward off microbial or insect attack upon plants, animals, and human artifacts. Others, however, believe that the natural world should be regarded as inviolable and that human intrusion should be minimized. Associated with this group are those who fear that technological interference with life forms or release of genetically engineered organisms could result in environmental catastrophies. Illustrative of this concern was the lawsuit of Jeremy Rifkin to prevent the experimental release of genetically altered bacteria on leafy vegetables as a protection against frost damage and a 1985 petition to Congress by the Environmental Policy Institute for a moratorium on deliberate release of genetically engineered organisms until the National Academy of Sciences completed a two-year study of the possible effects of such tests.[10] Conversely, some environmentally concerned persons advocate the use of biotechnology to control insects regarded as noxious in preference to environmentally damaging pesticides.

Few persons fall into only one of these six-fold areas of concern. The categories outlined here, with their diverse and adversary characteristics, have been presented to illustrate the formidable task of formulating coherent, consistent, and generally acceptable policies for biotechnology. For elective legislative representatives or public executives, the development of policies for biotechnical innovation approximates a no-win circumstance. It is therefore hardly surprising that governmental bodies have been reluctant to press for equitable methods to enhance the probability that biotechnological innovations will do more good than harm. Their constituents disagree over what is "good." Even so, opinion surveys reported in 1985 indicated widespread support in principle for research and development in biotechnology."[11]

Policy and Control

At this point a reader might expect to find a description of how the government of the United States and perhaps other national governments are organized to deal with biotechnological issues. Some reference might be made to what agencies have been established internationally to consider biotechnical problems. The following comments on these organizational topics will be brief for one good reason. Present arrangements are largely the results of ad hoc and tentative responses to the more salient issues. Moreover, there are detailed and accessible accounts of various policy controversies and regulatory efforts, some of which are referenced in the notes to this chapter. The topic falls into the larger category of public regulation of scientific research.[12]

In the Congress of the United States since 1970, nearly a dozen bills to regulate biotechnology have been introduced, but to date none have become law.[13] Nevertheless, regulatory action has been initiated at the administrative level. The House of Representatives Committee on Science and Technology has been the major source of policy initiatives. In a December 1986 report to the House Committee on Science and Technology, its Subcommittee on Investigations and Oversight set out the following summary of the course of regulatory action for biotechnology (in the more narrow sense) during the preceding decade:

> The federal regulation of biotechnology is now in its third stage. The first stage began with the "Guidelines for Research Involving Recombinant DNA Molecules" (NIH Guidelines) adopted by the National Institutes of Health (NIH) in 1976. These guidelines took a cautious stance, emphasizing containment standards and preliminary review by committees at each research institute of proposed experiments involving recombinant DNA

technology. The guidelines have been gradually relaxed as scientists obtained more information about the safety of the organisms involved.

The second stage of biotechnology regulation began on December 31, 1984, when the Office of Science and Technology Policy (OSTP) published a "Proposal for a Coordinated Framework for Regulation of Biotechnology," The purpose of the proposal was to "provide a concise index of U.S. laws related to biotechnology, to describe a scientific advisory mechanism for assessment of biotechnology issues, and to explain how the activities of the federal agencies in biotechnology will be coordinated."

The Proposed Coordinated Framework served as guidelines for the review of biotechnology products until June 26, 1986, when the Office of Science and Technology Policy published the "Coordinated Framework for Regulation of Biotechnology." This "Coordinated Framework" incorporated the comments received on the December 1984 Proposed Coordinated Framework and contained revised and expanded agency policy statements regarding the regulation of biotechnology products. This began the third and present stage of government regulation of biotechnology.[14]

Soon after its publication in June 1982 the administration's Proposed Coordinated Framework was challenged in a lawsuit filed in the United States District Court for the District of Columbia. On 22 December the case was dismissed by the court on grounds of "failure to present a justifiable case or controversy."[15]

The House of Representatives Subcommittee on Investigations and Oversight had meanwhile held hearings on the regulation of biotechnology including the Proposed Framework and had reviewed the concerns of the Congress "from the early days of genetic engineering to the present." The 1986 Subcommittee report summarized and referenced previous Congressional action on biotechnology policy. It thus seems evident that statutory action will be taken in the near future, possibly before publication of this book.

The Biotechnology Science Coordination Act of 1986 H.R. 4452 was introduced into the 99th Congress on 16 March 1986 and joint hearings were held by several House subcommittees on 4 and 5 June 1986. In the Senate the *Biosafety Act* S. 1967 was introduced on 17 December 1985 but did not receive a hearing during the 99th Congress. The House legislation would have established a Biotechnology Science Coordinating Committee within the Office of Science and Technology Policy in the Executive Office of the President. This committee would coordinate the review and assessment functions of the various federal agencies regarding information concerning genetically engineered organisms. In addition a Biotechnology Science Research Program would be established within the same office with joint industry-government funding to create a data base to support the development and regulation of biotechnology.[16]

Direct regulatory action, it appears, would remain with the "on line" operating agencies, notably the departments of Agriculture, Defense, Health and Human Services (including especially the Public Health Service, the National Institutes of Health, and the Food and Drug Administration), the Occupational Health and Safety Administration in the Department of Labor, and the Environmental Protection Agency.

It is clear from the aforementioned report that Congressional concern with biotechnology has related primarily to in vitro alteration of genetic material for the purpose of creating new genetic combinations or modifications. The scope of biotechnology, as treated in this book, is far broader. It seems unlikely that any single statute could encompass the entire field of biotechnological research or applications and there may be no need for such comprehensive legislation. Prevision of possible effects of significance and prevention of misapplied knowledge and technique are the principle points for concern, or would be in a society in which there was a high degree of consensus regarding rights, obligations, and values. But this is not the kind of society that today characterizes the technologically advancing nations of the world. From a policy perspective uncertainties regarding benefits and risks and the limited number of possible options for addressing them, complicate the business of policymaking. As federal judge David L. Bazelon has observed:

> If agencies and legislatures fail to address the risks of biotechnology—either to deter them or compensate their victims—the courts will likely do so, perhaps even more vigorously, in deciding private lawsuits. . . .
>
> Regulation through the common law has many drawbacks. It has a substantial impact on science, technology, and the economy generally. It "regulates" and constrains just as surely as an agency does. It can cause researchers to follow a variety of unnecessary practices, simply to avoid lawsuits. It also imposes extremely high costs in damages, insurance, and attorneys' fees. Moreover, judicial regulation cannot provide the consistency, rationality, or political responsiveness offered by a consciously designed and clearly articulated legislative solution. A courtroom is not the place to decide such complex and controversial issues of fact and policy. And judges are not the appropriate persons to decide them.
>
> Although the appropriate role for the courts in the regulation of biotechnology is limited, it is nevertheless important. It consists principally of judicial oversight of actions by administrative agencies.[17]

Biocracy, or the power of verified biological knowledge, may ultimately create a more consensual point of view than prevails today. Unless and until this happens public policy for the life sciences and biotechnology will remain an area of political contention. Factionalized and polarized politics seem likely to accompany advances in biotechnical capabilities that impinge upon moral values and economic interests. Prospects for a rational biopolicy ought therefore to be of concern to anyone who believes that in human self-knowledge and understanding of the biological factors in human behavior lies the best hope of humanity in preventing possibly irretrievable self-harm.

Is Rational Policy Possible?

The foregoing discussion of options, benefits, and risks makes evident the difficulty of formulating public policies to guide the development and application of biotechnology. There are few areas of public policy in which opinions are more divided and emotions as fervent. Under these circumstances several consequences follow. Legislative committees cannot find a consensual basis for public laws relating to biotechnology; governmental and private funding agencies are generally cautious in assisting policy-related research, especially on the more sensitive issues; the American inclination toward policymaking through lawsuits finds a wide field for adversarial action.

Unlike most public issues, the moral aspects of biotechnology have aroused religious concerns of an uncompromising character. Religious organizations have taken official positions on a variety of biotechnical applications, largely relating to human reproduction. And in many of these instances official pronouncements have been opposed by significant numbers of communicants and some members of the clergy. In March 1987 the condemnation by the Roman Catholic Congregation of the Doctriine of a number of biotechnical reproductive innovations, including an appeal to governments to prohibit their use, has placed this issue squarely in the arena of politics.[18]

Political organizations and movements have adopted policies that in some cases advocate biotechnological innovations and in other cases oppose them. For example both Marxists and free enterprisers find their ranks divided on various biology-related matters, with the curious result that the far "left" and far "right" find themselves in accord on issues such as overpopulation and in opposition to birth control through contraception but for dissimilar reasons. In brief, biotechnology seems an uncertain prospect for rationality in public policymaking.

It is therefore hardly surprising that public policy development for biotechnology has been halting and disjunctive, relating primarily to control of scientific research procedures. Inaction may be the wiser course, however, when opinion is divided and the rational course of action unclear. With few notable exceptions such as the Nova educational telecasts, the communications media have served the public poorly on biology-related issues. Characteristically, the media have emphasized adversarial and sensational aspects, contributing little to public understanding of the substantive issues or their implications. Thus the constituency for biopolicy is divided and, to a large extent, confused.

Given these circumstances, there is an obvious possibility that regulatory action relating to biotechnology might do more harm than good. Advancement in the life sciences might be retarded by governmental standards and regulations that crippled research and broadened the already wide field of litigation. Where political conditions suggest that governmental compromises might lead to mischievious results one might argue that the wisest public policy for biotechnology would be no policy, relying on public measures already in place to protect the health and safety of the population. But although this theoretical option has a superficial color of rationality, there are countervailing reasons to doubt its ultimate practicality.

Foremost among these reasons is the ramifying power of many biotechnologies. Life is dynamic and many altered life forms and processes affect other aspects of life in chain-like reactions. One need not advocate regulation of biotechnical innovation to argue that the implications of powerful new techniques should be studied, their possible consequences conjectured and taken into account. A capability for forecasting the courses and consequences of biotechnological innovation would be advantageous if for no other reason than to allow timely and nondisruptive adjustment to possible second- and third-order effects. To forecast is not necessarily to predict. There are risks in taking action prematurely to accomodate or prevent presumed consequences. Yet having tacitly assumed a responsibility for life-altering technologies, society, through its designated representatives, can hardly escape the additional responsibility of dealing with whatever consequences impact significantly on its future.

A second reason for doubting that a policy of governmental nonaction will prove feasible is the international character of science and technology. No nation alone can fully control the effects that biotechnology may produce even within its national borders. Restrictions on research and development in one country would cause determined investigators to carry on their work in other countries. Efforts to bar techniques that were regarded by many people as highly advantagious would impose an added burden of policing. Moreover, biotechnical developments affecting agriculture, medicine, and the economy generally could give an innovating country a competitive advantage over those nations that adopted restrictive policies. United Nations organizations and multinational corporations could promote the dissemination of biotechnical methods and products to the discomfiture of public officials of the restricting country. Recognizing the importance of biotechnology the

United Nations Industrial Development Organization (a specialized agency of 137 members since 1 January 1987) has established the International Centre for Genetic Engineering and Biotechnology with locations in New Delhi, India, and Trieste, Italy, and with an associated network of institutions in other countries.[19]

A third reason for questioning the feasibility of a no-action policy is the high probability of social acceptance of techniques found to be advantageous. Unless somehow inhibited to the contrary, people tend to seek out ways of doing things that serve their preferences or desires. Biotechnology creates opportunities and possibilities that permit the questioning of previously seldom-questioned moral dogma, but also could be exploited in fraudulent or harmful ways. The morality of bringing unwanted children into the world is now an issue in controversy. Recourse to biotechnical practices that are increasingly regarded as "good" and "just" will put traditional morality under stress. This in turn will lead to controversy as conventional moralists attempt a rear-guard action against the new ideas and practices. More pertinent may be action, similar to that taken in food and drug laws, to protect the public from physical harm.

Possibilities for rational policy for biotechnology might be significantly enhanced if the social issues, ethics, and political implications associated with its development received systematic study congruent with its dimensions. As of the mid-1980s no such systematic consideration had been undertaken on a scale commensurate with the subject. Social sciences concerned with public policy were generally resistant to an area of inquiry that by implication suggested that there might be linkages between the biological and social sciences that ought to be investigated. Sociologists and political scientists who had been fighting social Darwinism and abuses of eugenics have been understandably suspicious of efforts to reintroduce biological concepts into social philosophy. Biopolicy studies do not imply the uncritical injection of biological theories and findings into social science, but the risk is suspected nonetheless. Having ejected "biologism" from the front door of the discipline, political scientists have been on guard to see that it does not reenter through the back door of biopolitics.

Good science, however, has been built, in part, through learning from past mistakes. As the weight of evidence of the social relevance of advances in the life sciences has increased, the social sciences slowly and reluctantly have begun to recognize the legitimacy and even the necessity of attention to the biological "revolution." The first chapter in this book, "Biopolitics: Science, Ethics, and Public Policy," was published initially in 1964 in the *Yale Review* and, it seems safe to say, would not have been accepted by any social science journal at that time.

Fifteen years later interest among political scientists led to the formation in 1980 of the Association for Politics and the Life Sciences. In 1986 the association obtained status as an official section of the American Political Science Association.

In April 1987 a report entitled *Biotechnology, Public Policy, and the Social Sciences: Critical Needs in Teaching and Research* was prepared at the Institute for Advanced Study at Indiana University, by a team of four political scientists in consultation with reviewers from the biological sciences and humanities. The report was addressed to government, philanthropic foundations, colleges and universities, and the social science disciplines. It recommended a commitment to interdisciplinary teaching and research on public policy and the life sciences. Whatever its immediate effect, the report represents progress toward recognition of the political significance of biotechnology and the life sciences generally.

If the foregoing analysis is correct, the ultimate emergence of a rational policy for biotechnology and the life sciences seems possible, at least in the scientifically advanced countries. International policy and agreement also seem possible because of the worldwide effects of biotechnical developments in agriculture, medicine, and human reproductive technologies. The possible is, of course, not necessarily the probable. One could easily underestimate the forces of resistance to the changes in belief and practice that are implicit in many life sciences advances. The depth and breadth of social and political conflict that may result from public decisions on biotechnological options cannot be foreseen. But the incidence and intensity of conflict could be alleviated by unprejudiced interdisciplinary analysis of the issues and the alternatives for dealing with them. Knowledge can be employed as a force for rationality.

But will it? The greatest danger in the new biotechnology is its potential for use in warfare.[20] Biological and biochemical technologies have been developed for military purposes and are presently available in the arsenals of nations. How many national governments have biological warfare capabilities is conjectural, but this is a form of weaponry that poor nations can afford and there are many poor nations. Risk of back-contamination may have deterred use of biological agents in war, but a small nation that could immunize its citizens from a lethal virus might deploy it against an invader. The boundary between biological and chemical agents is not distinct, and both may be present in biochemical military technology. Here indeed is dangerous knowledge, but dangerous because of the mind-set of the people who conceived it and might deploy it. Concerned citizens protest the stock-piling of biochemical weapons but might more rationally work to remove the state of mind that makes weaponry appear needed and rational. Nicholas Berdyaev observed that:

> Modern thinkers like to talk of the "frontier situation" of man, of the dangers besetting him on every side (Tillich, Jaspers). And there is nothing surprising in this, since history evidently does not undertake to give man any guarantees of existence or offer any protection for his life.[21]

One can hardly escape the conclusion that the balance between the benefits and risks of biotechnology will be tipped by human purpose and moral insight. Technology is not neutral, but neither is it purposive. The options in its development and use are human choices. The choice may be between a superior humanity or self-destruction. How the new biotechnological powers will be used cannot be foreseen because unpredictable human behavior is the critical factor. In the course of time, the judgment of history will reveal the verdict.

Notes

1. For definitions and comment relevant to biotechnology as considered in this paper, see Arthur D. Little, Inc., *Study of Federal Biotechnology Policy Issues: A Report* (Washington, D.C.: National Science Foundation, 12 July 1984), 1–5; Henry I. Miller and Frank E. Young, "Biotechnology: A Scientific Term in Name Only," *Wall Street Journal*, 13 January 1987, 28; Sandra Panem, ed., *Biotechnology: Implications for Public Policy* (Washington, D.C.: Brookings Institution, 1985); U.S. General Accounting Office, *The U.S. Department of Agriculture's Biotechnology Research Efforts* (Washington, D.C.: U.S. Government Printing Office GAO/RCED-86-39 BR, October 1985); U.S. Office of Technology Assessment, *Commercial Biotechnology: An International Analysis* (Washington, D.C.: Government Printing Office OTA-BA-218, 1984); Steve Prentis, *Biotechnology: A New Industrial Revolution* (New York: George Braziller, 1984); and Steve Olson, *Biotechnology: An Industry Comes of Age* (Washington, D.C.: National Academy Press, 1986).

2. Peter J. Ucko and G. W. Dimbleby, *The Domestication and Exploitation of Plants and Animals* (Chicago: Aldine, 1969); and F. E. Zeuner, *A History of Domestic Animals* (London: Hutchinson, 1963).

3. E.g., Pamela Sargent, *Bio-Futures: Science Fiction Stories about Biological Metamorphosis* (New York: Random House, 1976).

4. Lawrence J. Henderson, *The Fitness of the Environment: An Inquiry into the Biological Significance of the Properties of Matter* (New York: Macmillan, 1913), 266. For a contemporary account of environmental chemistry updating Henderson, see R. A. Bailey, et al., *Chemistry of the Environment* (New York: Academic Press, 1979).

5. There is a large and growing literature on risk assessment and public policy. See W. W. Lowrance, *Of Acceptable Risk: Science and the Determination of Safety* (Los Altos, California: William Kaufmann, 1976); Sandra Panem, ed., *Public Policy, Science, and Environmental Risk* (Washington, D.C.: Brookings Institution, 1983); Allen Jones and Walter F. Bodmer, *Our Future Inheritance: Choice or Chance?* (Oxford: Oxford University Press, 1974); Susan G. Hadden

and Jared Hazelton, "Public Policies toward Risk," *Policy Studies Journal* 9 (Autumn 1980): 110–18; David Okrent, "Comment on Societal Risk," *Science* 208 (25 April 1980): 372–75; David L. Bazelon, "Risk and Responsibility-Risk Regulation: A Problem for Democracy in the Technological Age," *Science* 205 (20 July 1979): 277–80. See also the journal *Risk Analysis*, 1981; Robert H. Blank, "The Challenge of Emergent Public Policy Issues in Genetic Counseling," (address before the National Society of Genetic Counselors, Philadelphia, 31 October 1986 [unpublished as of mid-1987]); U.S. Office of Technology Assessment, *Impacts of Applied Genetics: Micro-Organisms, Plants, and Animals* (Washington, D.C.: OTA, April 1981), and *Genetic Screening in the Workplace* (Washington, D.C.: OTA, 1983).

6. See Robert H. Blank, *The Political Implications of Human Genetic Technology* (Boulder, Colorado: Westview, 1981).

7. On the recombinant DNA issue, from a large body of published work, see S. Krimsky, *Genetic Alchemy: A Social History of the DNA Controversy* (Cambridge, Massachusetts: MIT Press, 1983); J. Morgan and W. J. Whelan, eds., *Recombinant DNA and Genetic Experimentation* (Oxford: Pergamon Press, 1979); R. A. Zilinskas and B. K. Zimmerman, eds., *The Gene Splicing Wars: Reflections on the Recombinant DNA Controversy* (New York: Macmillan, 1986).

8. Robert H. Blank, *Rationing Medicine* (New York: Columbia University Press, 1988).

9. E.g., Stephen Kreider Yoder, "Japan Jumps Ahead in Use of Biosensors: Devices Unite Organic Substances, Electronics," *Wall Street Journal*, 23 December 1986, 22.

10. See Keith Schneider, "Biotech's Stalled Revolution," *New York Times Magazine*, 16 November 1986, 42 ff. See also note 14 below.

11. See "Biotechnology Gets Good Marks in Survey," *Chemical and Engineering Now* 63 (26 August 1986): 16; also "Polling Biotechnology Attitudes Within and Without," *ASM News* 51 (November 1985): 1.

12. See Jack Doyle, "Biotechnology: Regulating the Risks, Data Gaps, and Needed Research," remarks presented at the Institute for Sustainable Agriculture Symposium, "Implications of Biotechnology for Sustainable Forming Systems," Washington, D.C., 4 March 1986; also Elliot S. Gershon, "Should Science Be Stopped? The Case of Recombinant DNA Research," *The Public Interest* 71 (Spring 1983): 3–16; Malcolm L. Goggin, "The Life Sciences and the Public: Is Science Too Important To Be Left to the Scientists?" *Politics and the Life Sciences* 3 (August 1984): 28–40, and review commentaries and references, 40–75; and Leonard A. Cole, *Politics and the Restraint of Science* (Totowa, New Jersey: Rowman and Allenheld, 1983).

13. In the 99th Congress (1986) three bills (H.R. 4452, H.R. 5271, and S. 1967) were introduced; none passed. See also Marjorie Sun, "Biotechnology's Regulatory Tangle," *Science* 225 (17 August 1984): 697–98; Albert Gore, Jr., "A Congressional Perspective," in *Biotechnology: Implications for Public Policy*, ed. Sandra Panem (Washington, D.C.: Brookings Institution, 1985); and Odelia Funke, "Biopolitics and Public Policy: Controlling Biotechnology," *PS* (Winter 1985): 69–77.

14. House Committee on Science and Technology, Subcommitte on Investigation and Oversight, *Issues in the Federal Regulations of Biotechnology: From Research to Release,* 99th Cong., 2d sess., December 1986, Committee Print, 2. See this entire report for coverage of federal action relating to biotechnology.

15. See Mark Crawford, "Court Rejects Rifkin in Biotech Cases," *Science* 235 (9 January 1987): 159; and "Rifkin Sacked," editorial, *Wall Street Journal,* 26 January 1987. See also Marilyn Chase, "Jeremy Rifkin Usually Infuriates—And Often Beats—Biotech Industry," *Wall Street Journal,* 2 May 1986, 15. For earlier litigation on this issue, see Gerry J. Elman, "Environmentalists Ask Court to Halt Testing of Recombinant Plant Symbiotes in the Field," *Genetic Engineering News* 3 (November–December 1983): 1.

16. *The Biotechnology Science Coordination Act of 1986,* Hearings before the House Subcommittee on Natural Resources, Agriculture Research, and Environment and the Subcommittee on Science, Research and Technology of the Committee on Science and Technology, 99th Cong., 1st sess., 4–5 June 1986, no. 150. These hearings provide a valuable source of information on the status of federal policy regarding biotechnology and the positions toward it taken by a range of interested groups—scientific, industrial, environmental, and legal.

17. David L. Bazelon, "Governing Technology: Values, Choices, and Scientific Progress," *Technology in Society* 5, no. 1 (1983): 19.

18. "The Conjugal Union Must Take Place," excerpts from the Catholic Church's "Instruction on Respect for Human Life in Its Origin and on the Dignity of Procreation," *Washington Post,* 11 March 1987, A4; Edward M. Yoder, "Procreation: The Vatican vs. Reality," *Washington Post,* 17 March 1987, A17; Joseph Carey, "Procreation, Science, and Sin," *U.S. News & World Report,* 23 March 1987, 10.

19. *UNIDO Newsletter* 214 (February 1986) and 216 (April 1986); "Three Giant Steps for International Gen-Tech Center: $18M Interim Program Finalized," UNIDO News Release (New York: UNIDO, 22 October 1986); also Dermot A. O'Sullivan, "International Biotechnology Center Takes Shape," *Chemical and Engineering News* 64 (13 October 1986): 21 ff. For background, see Peter Dorfmann, "International Biotechnology Center Declared to Exist . . ." *Genetic Engineering News* 3 (November–December 1983): 1; Burke K. Zimmerman, "India and Italy Force Compromise on International Biotech Center Site," *Genetic Engineering News* 4 (May–June 1984): 3; David Dickson, "UNIDO Hopes for Biotechnology Center," *Science* 221 (30 September 1983): 1351–53; and the following publications of UNIDO: *The Establishment of an International Center for Genetic Engineering and Biotechnology (ICGEB), Report of a Group of Experts,* UNIDO/15.254, distr. ltd.; *Elements of Some National Policies for Bio-technology: Note by the Secretariat of UNIDO,* UNIDO/15.270/Rev.1., distr. ltd.

20. See Raymond A. Zilinskas, "Recombinant DNA Research and Biological Warfare," in *The Gene-Splicing Wars: Reflections on the DNA Controversy,* extensive bibliographical notes. See also "The Soviet Biowar," *Wall Street Journal,* November 1986, editorial; Jeffrey R. Smith, "New Army Biowarefare Lab Raises Concern,"

Science 226 (7 December 1984): 1176; and Philip M. Boffey, "Pentagon Fears U.S. Biotechnology Could Be Used by Russia as Weapon," *International Herald Tribune*, 29 May 1984.

21. Nicholas Berdyaev, *The Fate of Man in the Modern World*, trans. Donald A Lowrie (London: Student Christian Movement Press, 1935), 22–23.

9
International Consequences of Applied Biology

We may understand applied biology in two senses, one general, the other special. The former refers to a broad range of uses of biological knowledge, informational and conceptual as well as technical.[1] The special meaning, more common in scientific usage, applies the term *biotechnology* to the employment of living material (e.g., microorganisms or genes) as tools. In the literature of microbiology and genetics, biotechnology may be restricted to the special meaning. But outside the laboratory in the public domain where life is affected by technologies indirectly utilizing biological knowledge or living materials the more general meaning of biotechnology is appropriate. Manipulation of life forms and processes requires assistance from a great number of physicochemical technologies. Narrowly defined distinctions seldom reflect what happens in the world of practical affairs.

The accelerating and often unprecedented nature of present-day advances in all aspects of applied biology is rapidly bringing about a situation that is unlike any that has previously been experienced. This situation can be stated simply, even though its implications are complex and not fully evident: Biotechnological innovation has now attained a potential for social and ecological impact that cannot be managed wisely or effectively in the absence of some system of universal international responsibility.

In a world of nations, international responsibility implies national responsibility. In matters affecting the course of life on earth, these two domains—national and international—cannot realistically be separated. International interactions and effects attributable to biotechnology originate within national boundaries and, crossing them, impact upon other nations. National successes, failures, or irresponsibilities relating to biotechnical developments—including in some instances their rejection—can seldom be contained. International efforts to control the spread of diseases communicable among animals and plants has long demonstrated the permeability of national frontiers. Today, more than ever, there is a growing body of knowledge and techniques whose dispersion will be as difficult to control as most forms of infectious disease. And so applied biology and more particularly biotechnology has assumed a place of growing recognition and importance on the agenda of international affairs.

International Context of Applied Biology

Applied biology and especially biotechnology demonstrate a growing potential to affect the duration and characteristics of almost every form of life and to create new life forms. Biotechnology, in particular, may enhance the survivability of some species while controlling or exterminating others. The reality and potentiality of this capability has not as yet borne in upon the public consciousness. Unlike nuclear energy which tends to be viewed with general apprehension, biotechnology arouses public concern primarily in relation to specific issues. Different publics are involved in different developments in relation, for example, to birth control techniques, recombinant DNA, experiments with animals (especially humans), patented germ plasm, and biological warfare. This fractionated perspective would be of less consequence but for the tendency of changes in the living world to have ramifying, synergistic, and often unforeseen effects. Difficult as control of the atom may be, living processes are characterized by a variability, subtlety, complexity, and unpredictability that makes the control of biotechnology more difficult and control methods more diverse.

The politics of applied biology cannot be discussed without some deliberate generalization and simplification. We have used the term biotechnology in a more general sense than some life scientists would prefer because in the actual application of biological knowledge to human purposes closely defined cause-effect situations are exceptional and specialized biotechnologies are often applied in association with other science-based techniques, usually chemical or mechanical. Neither the public nor its political representatives number many individuals who possess detailed understanding in the life sciences; few are able to appreciate refined distinctions. Mankind's future could be made more promising by a worldwide increase in this understanding, extending beyond the technologically advanced countries. Biotechnical developments in these countries have already affected the entire world and may continue to do so at an accelerating rate.

Thus far the effects of biotechnology are most apparent in two interrelating subject areas, one of which primarily concerns animals, the other relating mostly to plants. Both involve microorganismic life.

With the exception of biological warfare, the objective of biotechnology in both areas of application has been the perceived welfare of people, but some benefits of biological innovation have been accompanied by disturbing and even disrupting side effects. The most widely publicized achievements in biotechnology have been in agriculture and medicine, although reproduction-related technologies (e.g., frozen embryos, genetic screening, and in vitro fertilization) have been widely publicized.

As noted in the preceding chapter, applications of biotechnology to animals have been extended not only to humans through medicine, nutrition, and reproductive techniques, but in a broader sense extensively and for centuries to domesticated animals and more recently to wild animals.[2] Biological and chemical interventions in agriculture through more conventional technologies of hybridization, and application of fertilizers and pesticides have been supplemented by "genetic engineering" of new characteristics and resistances in selected plants. Less attention has been attracted by advances in silviculture, in the growing commerce in patented germ plasm for cereal grains, or in techniques for ascertaining the take-up from the soil by plants of toxic substances which may be transmitted up the food chain to humans. Of particular significance is the development of nitrogen fixing and pest resistant plants that could reduce, possibly eliminate, international commerce in fertilizers and pesticides.

International aspects of biotechnology are not wholly novel and are conditioned by circumstances that characterize many other relationships among nations. Foremost among these is the uneven distribution of bioscientific and biotechnical resources around the globe. Present technoscientific capabilities (not necessarily potentialities) are concentrated in a few highly developed countries (e.g., in the United States, United Kingdom, France, West Germany, Switzerland, and Japan). This observation is not a criticism, but is recognition of a circumstance that affects international relationships. Applications of biotechnology, although concentrated in the more highly developed countries, are now extended to the less developed states. These countries seldom possess the scientific or technoeconomic infrastructure to evaluate or manage imported biotechnology. Thus, to the extent that they become dependent upon external resources for capabilities in important sectors of their economies such as agriculture, medicine, nutrition, or reproduction technology, they may correspondingly lose control over some aspects of their domestic policies. The "green revolution" and genetically engineered hybrids have sociopolitical, economic, and ecological effects that suggest the need for care and caution in their adoption.[3]

More significant politically is the perception of dependence and loss of control. The extent of governmental control over uses of biotechnology is limited even in the most highly organized states, not excluding the most centralized and authoritarian. Governments have not enjoyed notable success in promoting fecundity, suppressing abortion, curbing alcohol and drug abuse, establishing sustainable agricultural practices, or containing the spread of technical knowledge. Where people see available biotechnologies as serving their personal preferences they tend to adopt them regardless of official policies. Where government policies and perceived advantages coincide, adoption of techniques may be rapid and widespread.

Of course, some people enjoy wider options or access than others to some biotechnologies which can be controlled by political or economic means. In developing countries, for example, large landowners, corporate enterprises (national and multinational), and government agencies—notably military—have greater latitude in adoption of biotechnical innovations than most people. Political orators in developing countries have in the same speech denounced technoscientific imperialism but also demanded transfer of science-based technologies, as if this were merely a matter of acquiring instruments, methodologies, and legal rights. This is a logical reaction to perceived dependency and a desire for policy control. But like most political appeals it oversimplifies a complex set of interrelationships and overlooks the crucial factor of receptive capability in any transfer of technology.

At least three sets of circumstances complicate the ways in which biotechnology affects international relations. First, although not always conceded by governments, is the readiness of a country to effectively utilize technological innovation. Second is the multiplicity of channels through which various technological innovations may be introduced. Third is the frequent inability of governments to monitor or control the uses of biotechnology as, for example, in the importation, distribution, and use of pesticides and pharmaceuticals, especially in countries at low levels of technoscientific development. In centralized, planned, and policed economies, the spread of new ideas or techniques can be restricted. Yet even in centrally controlled police states, the presence of a new and advantageous foreign technology may present a troublesome problem of control. Cybernetics, for example, was initially resisted in the USSR until Soviet scientists at last convinced the government that an advantageous technoscientific innovation should not be proscribed for theoretical, ideological reasons. Imported contraceptive technologies have been prohibited in some countries.

Uneven receptivity among nations is a handicap to general international agreement regarding the transfer of biotechnology. Throughout much of the world today, conditions are not favorable for the informed and prudent application of biotechnology. In the least developed countries illiteracy, resistant tradition, poverty, and often civil disturbance prevent even a governing elite from effectively introducing innovation. For example, in the Iran of the Ayatollah Khomeini, receptivity to imported biotechnology has been extremely limited. Even before the Ayatollah's xenophobic regime, it was reported that efforts of international technical assistance experts to introduce improved poultry breeding in Iran incurred the suspicion and the hostility of peasant farmers. It appears that arch-conservative religious mentors suggested that experiments with domestic hens could be a prelude to manipulating the sex life of the village women.[4]

Whatever the authenticity of this incident, and there have been others similar in effect, it is clear the biotechnology frequently touches upon attitudes and mores of intense personal concern. Absence or deficiency of a scientific infrastructure within many countries necessitates reliance upon foreign experts who may not be received with confidence by local populations and who are themselves poorly prepared to relate to indigenous cultures. Governments, moreover, are often ill-equipped to evaluate, monitor, or police imported technologies whether agricultural, pharmaceutical, genetic, or medical. One consequence of a sense of vulnerability among leaderships in less developed countries is to turn to international institutions for protective assistance.

A difficulty confronting the assistance-seeking country is the multiplicity of channels and forms through which biotechnology may enter. Radio, television, and the press carry international news dispatches regarding artificial hearts, organ transplants, new forms of birth control, "test-tube" babies, and controversies over pesticides and experiments with recombinant DNA. Whereas initially only a small elite may receive this news, the "global village" effect of present day telecommunications via satellite implies a widening dispersion of information. Even where the news media are tightly controlled, word of mouth by diplomats, scientists, medical missionaries, and other persons traveling abroad spreads the news.

The principal means for internationalizing and disseminating biotechnology are international organizations—corporate, intergovernmental, scientific, and philanthropic. The multiplicity and diversity of these organizations in relation to biotechnology has made it a subject for international concern that even now may be generating new intergovernmental arrangements, and perhaps new principles of international law.

An important part of the international transfer of biotechnology, or indeed of any international transfer of technique, is the background of information or experience against which contemporary transactions are evaluated. Biological information or concepts may prepare the way for technological application. Publicized and sometimes analyzed events have influenced national and international policies on certain biology-related matters. Cases in point are cumulative experiences with introductions of plants and animals exotic to a new environment. Rabbits in Australia and water hyacinths and starlings in America are familiar examples of the explosive multiplication of species in alien environments. Serious economic consequences of inadvertent as well as deliberate importations of exotic organisms (i.e., the chestnut blight, elm disease fungus, and Mediterranean fruit fly in America) have caused governments to negotiate international treaties to protect against disastrous ecological disruptions.[5] Such experience has also caused popular apprehension over the experimental release of ostensibly harmless microorganisms into the environment.

In more controversial areas, sequential events have been instructive. Explosion of human population growth following the World Health Organization's antimalaria campaign and the increase of schistosomiasis associated with large-scale water resource development are two instances in which well-intentioned interposition into natural processes was unaccompanied by measures to ward off unwanted (and largely unanticipated) consequences. Release of genetically altered microorganisms into the environment, even under strict cautionary controls, arouses concern that has led to litigation, as in efforts to prevent the field-testing of bacteria to reduce risk of frost damage to sensitive crops.[6]

Legal obstruction of field testing in technologically advanced countries may be a factor in causing experimental work to be done in countries where few rules apply. For example, The Wistar Institute of Philadelphia in collaboration with the Pan American Health Organization was reported to have field tested genetically engineered living rabies vaccine in Argentina without the knowledge of the Argentine government. *The United States Coordinated Framework for the Regulation of Biotechnology,* published 26 June 1986, does not prohibit or control American companies or research laboratories from testing genetically engineered products in other countries. The experiment in Argentina was conducted at an agricultural station but the government had no policies or rules applicable to such a case.[7]

The use of animals for experimental purposes has been a subject of some concern, especially when international trade in species is involved. Biotechnology is not involved in all experiments with animals, but may have value in the perpetuation of endangered species through reproduction techniques and biological control of damaging insects. The idea of unique humanity has been so deeply ingrained in modern cultures that it has been very difficult to generalize in popular discourse about behaviors common to all animal life. Students of animal behavior (i.e., ethologists) have identified behavioral patterns common to many different groups of animals, including humans. Many animals, especially some closely resembling humans in certain respects (e.g., monkeys and chimpanzees) have been used extensively for experimental purposes in biology and medicine. Most discussions of the impact of biotechnology focus upon human life—life of greatest interest to most humans. That is the focus here; but it is nonetheless true that impacts of bioscience and technology are felt by all forms of animal life, both those domesticated and (in the perspective of humans) "wild." Some very specific issues (e.g., transmission of foot and mouth disease among domestic animals or export of endangered plants or animals) may escalate into diplomatic controversies among nations.

The objectives of the discussion here will be best served, however, by the selective examination of categorical areas in which biotechnology as applied biology has become a factor in international affairs. Subsidiary issues, such as the infant formula controversy, are introduced primarily to illustrate more inclusive aspects of international biopolitics. Within the larger category of health, nutrition, and reproduction an obvious area of international relations affected by biotechnology in the broad or general sense is human health.

Health and Nutrition

To understand international aspects of the biotechnology of health, it is necessary to recognize the difference between maintenance of basic health services (primary health care), disease prevention, and the treatment of disease (therapeutic medicine). An area of activity involving both primary health care and medicine is the correcting or compensating of disabilities through bionic devices. These distinctions are important in international relations for several reasons. First, advancements in biomedical science and technology are concentrated in Europe, North America, and Japan, whereas a rapidly increasing percentage of the world's peoples live in countries lacking the institutional or economic means to obtain more than marginal benefit from these advancements. Second, the concept of a human "right" to health has permeated political ideologies and agendas and is now generally acknowledged by governments aspiring to create social service or welfare states.

This difference in the ability to provide health care services has several consequences that are reflected in international politics. For example, many national governments are unable to deliver upon their promises for universal health care. For this inability there may be many reasons, but the cost of advanced biomedical services is a major factor almost everywhere. For millions of peoples in Africa, Asia, and Latin

America, prevention of disease, or reduction of its incidence is the most that can realistically be expected. Thus so-called primary health care (e.g., including sanitation, nutrition, and preventive medicine) has become the major focus for public health policy in the less developed countries. Environmental conditions have been found to be major factors in health and disease everywhere, but are especially relevant in countries in which housing and sanitation are poor, water supplies are contaminated, and the living habits of people expose or predispose them to a variety of health risks. These environmental conditions are not readily remedied, and there is, accordingly, great interest in finding biotechnical means that will provide preventions or cures.[8]

The private commercial development of biotechnology has tended to channel its applications to the developed countries where returns of investment are promising. There are, meanwhile, large areas of potential medical application in developing countries, but financial and organizational problems are deterrents. However, a new Geographic Medicine Division at Stanford University Medical School is directing its efforts toward applying molecular and genetic medical techniques to diseases epidemic in the Third World. The Pasteur Institute in France has concluded an agreement to provide hepatitis B vaccine to Taiwan. Of greater significance for the long-run is the establishment of the International Biotechnology Center, presently to be discussed.

The more- and less-developed countries have, at present, distinctly different emphases in their policies for health. The differences occur not so much in concepts of what constitutes human health as in the emphasis given to alternative ways to achieve or maintain it. In Western societies, even to some extent under socialist regimes, maintenance of health (as distinguished from medical services) is largely treated as an individual personal responsibility. Especially in the United States, health is generally regarded as a personal affair except where containment of contagion, or research on widely endemic disease (e.g., cancer, AIDS, and cardiovascular disease) requires collective action.

Primary health care or preventive medicine may reduce the incidence of severity of disease, but there are many forms of illness that will not or cannot now be prevented. A growing number of biomedical products, notably pharmaceuticals (i.e., "drugs") are exported from more to less scientifically advanced countries. This commerce is largely private, much of it through multinational drug companies and its reception and use in less-developed countries is characteristically poorly understood and poorly monitored. In these countries the restrictions and controls governing the release of many drugs are either absent or are ineptly administered. The same circumstances apply to international trade in biocides (pesticides and herbicides). The so-called Third World thus provides markets, both legal and illicit, for substances banned or supervised in the advanced countries. As awareness of this circumstance has grown, so has resentment and protest in both sending and receiving countries.[9]

Even when products are in themselves harmless, the conditions under which they may be used may cause health problems. The export of commercial formula for the feeding of infants in developing countries created a major controversy involving the United States and the World Health Organization that set off political and ethical reverberations that lasted for many months.[10] The problem here was lack of adequate sanitation in the use of the formula in villages with unprotected water supplies. Comparable problems have arisen with misuse of agricultural pesticides where needed provisions for caution were not taken or understood. These are not biotechnical issues, but they affect the international political climate in which biology-related issues are considered.

A wide range of products in international commerce from nuclear materials to pills, pesticides, diet supplements, and even treated wearing apparel have come to be classified as hazardous. Because concerned people in less-developed countries believe their own governments to be unable to prevent or control the influx of these substances, they have sought to internationalize the problem, primarily through the United Nations system. On 17 December 1982 the General Assembly of the United Nations by vote of 146 to 1 adopted a resolution on "Products Harmful to Health and the Environment."[11] The United States cast the single dissenting vote on the logical but unrealistic hypothesis that each country ought to be responsible for policing its own imports, and controlling substances sold in its domestic markets. This rationale was offered to explain the rescinding on 17 February 1981 by President Reagan of President Carter's Executive Order 12264 of 15 January 1981 "on federal policy regarding the export of banned or significantly restricted substances."[12] The official position was that it was inappropriate for the United States to extend its internal policies to other countries, and that American companies in their operations abroad should not be required to conform to standards and procedures not required by host countries. The export of genetically altered microorganisms for field testing in countries with no or minimal controls could become a point of international concern.

As environmental aspects of human health are better understood, commerce in potentially harmful substances is almost certain to be brought under national and international control. Conversely, the dissemination of benefical biotechnologies, medical and nutrition related, seems certain to be assisted through international arrangements, private as well as public. The Wistar Institute in the United States and the Pan

American Health Organization were party to field tests for rabies vaccine undertaken in Argentina in 1986. But experimental work with genetically altered organisms causes apprehension. Recognition of an incipient international problem has persuaded the national governments comprising the Organization for Economic Cooperation and Development (OECD) to adopt policy guidelines for the safe environmental release of genetically altered organisms resulting from recombinant DNA technology.

Primary health care and disease prevention will probably continue to receive the greater emphasis in developing countries although biotechnology may assist in important ways chiefly relating to nutrition and control of harmful microorganisms. To the extent that general international agreement is obtained for measuring the quality of human health, the political significance of section II of the Declaration of Alma-Ata becomes more than rhetoric. Adopted by the International Conference on Primary Health Care, 1978, and endorsed by the thirty-second World Health Assembly and the United Nations General Assembly at its thirty-fourth session, the Declaration states that: "The existing gross inequality in the health status of the people particularly between the developed and developing countries as well as within countries is politically, socially and economically unacceptable and is, therefore, of common concern to all countries."[13] The implication is plain. Verifiable knowledge of comparative levels of health and nutrition among countries becomes an activating force in international politics.

The interrelating character of life sciences and technologies is illustrated by nutrition, which not only links health with agriculture, but joins both to a range of economic and technological factors including techniques of food processing and transportation, public preferences and advertising, and priorities in uses of water, land, and labor. Nutrition relates to more than strictly agricultural considerations, presently to be examined. It involves not only what is grown (and not grown) as food, but, more to the point, to what happens to the food produced—to its nutritive quality, its protection from contamination, and to its distribution and associated costs.

Human malnutrition has become a political issue in addition to an object of moral, economic, and social concern. Numerous international nongovernmental organizations (e.g., Oxfam, Red Cross, religious organizations) mobilize private resources to redistribute food, but also urge governments to cooperate in the alleviation of hunger. Bioscience, specifically the nutritional aspect of physiology, has produced the knowledge and the nutritional standards that are recognized by the health professions, by national governments, and by international bodies such as the Food and Agriculture Organization (FAO) and the World Health Organization (WHO). Ascertaining responsibility for human malnutrition and determining what should be done about it and by whom, involves much more than mere factual evidence. Matters of judgment, moral, political, and technical are involved. The issue of malnutrition may be summarized with equal logic as "too little food" or as "too many mouths." Pollution of air and water enters into nutrition policy not only because of risks of contamination, but actual diminution of plant and animal life as a consequence of such phenomena as acidic precipitation and persistent pesticides.

Science, international organization, and communications technology combine to give specificity to nutrition as an international issue. An example of this is the publication of the WHO entitled *Measuring Change in Nutritional Status: Guidelines for Assessing the Nutritional Impact of Supplementary Feeding Programmes for Vulnerable Groups* (1983). The guidelines are intended to assist countries receiving food aid to identify the presence of nutritional changes in selected population groups benefiting from food supplements (infants, preschool children, and primary school children). The measurements, sampling, and data collection and analysis involved in the development of the guidelines are necessary to the effective application of advances in the life sciences, and when these techniques are applied to target populations in particular countries, these findings can influence the course of multinational and bilateral aid programs.

Population and Human Reproduction

Applied to humans, reproductive biotechnology has become one of the more sensitive biopolitical issues. Its linkages with other issues are numerous, although some are not readily apparent. This field of biotechnology continues to develop and eludes efforts by political and religious authorities to contain or control it.

The importance of reproduction technology has been greatly increased over large parts of the earth by the explosive growth of human populations that followed the successful introduction of infectious disease control and improvements in food supply. The WHO campaign against malaria illustrates the interrelating effects of medical and biotechnologies and shows why it may be dangerous to introduce new biomedical techniques without taking account of the range of potential consequences. Population dynamics is a complex phenomena and so a poor subject for a single cause explanation. Nevertheless, the failure to balance birth control with death control in the middle years of this century have left in many countries a legacy of overpopulation accompanied by poverty that has in turn given rise to serious social, economic, and political problems.

Reproduction technology is the most personal, intimate, and socially sensitive innovation that has emerged from the life sciences. There have been many traditional conception control techniques, antecedent to science-based reproduction or population control technologies. Reproduction biotechnology, however, involves more than contraception. Not only may human fertility be suppressed or increased; prenatal defects may be detected at fetal stages or may be predicted from the genetic history of possible parents. Genetic screening has not yet become a conventional technique and has not yet gained the acceptance accorded the so-called Wasserman test for venereal disease required by law in many states. Civil rights activists are prominent in objections to screening for the AIDS virus on grounds of invasion of privacy and discrimination against presumed carriers. Acceptance of genetic screening will however increase as its benefits are recognized. For example, screening for certain genetic defects in newborn infants is now required in at least eighteen of the United States and eight other countries.[14]

Explosive population growth has been a contentious issue in many countries and has complicated international relations. During the early stages of more-advanced societies, population growth is generally favored by religious, political, and military leaders and most economic interests. Contraceptive counseling and devices have often been prohibited. When, however, housing, health, and educational services cannot keep pace with ever-growing numbers, and when poverty, unemployment, migration of indigents, and social unrest become the visible consequences of too many people too fast, resistance to reproduction technologies sometimes begins to soften. Still, there are obstacles.

The equalitarian tendencies of modern social democracies and their assumptions in theory that political power and human numbers should be proportional encourages the growth of upward-aspiring ethnic groups. Their interest in reproduction control is to increase, not to reduce, their numbers. International transfer of reproduction technology was initially inhibited by moral or ideological objections in donor as well as receiving countries. President Dwight D. Eisenhower firmly opposed the inclusion of contraceptive technologies in the United States bilateral aid program. Yet, the blighting of individual human lives, the frustration of social welfare programs, attrition of the natural environment, and the deterioration of burgeoning cities became so unmistakably associated with excessive population growth that as early as 1946 a population council was established within the Economic and Social Council (ECOSOC) of the United Nations. Opinions within and among nations on the scope and content of population policies have differed sharply and continue to do so. Gradually, however, the matter of numbers and reproduction technology gained legitimacy as an issue.

In 1967, the United Nations Fund for Population Activities was established and an account of its initiation and its remarkable development has been provided by its longtime Executive Director, Rafael M. Salas of the Philippines.[15] In 1970, the U.N. General Assembly on the recommendation of ECOSOC designated 1974 as World Population Year, the principal event of which was an intergovernmental World Population Conference held 19–30 August in Bucharest, Romania. The debates at Bucharest have been described as intense and conflict laden; nevertheless, amid the rhetoric of nineteen resolutions, a number of recommendations were adopted that, in effect, pointed toward international issues relating to biotechnology. From that number the following three excerpts have been selected:

> Resolution XI: Regional Cooperation
>
> 3. *Recommends* the use of technological expertise to improve, maintain, or enhance the regional and international environment and prevent pollution to the greatest extent possible;
>
> Resolution XV: Population and Research
>
> 5. *Suggests* that the United Nations should support national and international programmes which deal specifically with biomedical research on fertility and its variations, both normal and pathological, and research on the interaction between fertility and demographic variables and socioeconomic factors;
>
> Resolution XVI: Population Policies
>
> *Recommends* . . . (e) that national sovereignty and human rights should be fully respected in any programme comprising experimentation regarding methods relating to fertility. Such methods should first be approved by the governments both of the country of origin and of the country in which they are used.[16]

Clearly in a world in which mere numbers of humans are equated with proportionate entitlements to resources, space, economic status, and political power, population policies and reproduction technology are inevitable items on the agenda of international affairs. The liberal and equalitarian principles proclaimed in the more-advanced countries puts them at a logical, even embarassing, disadvantage in altercations with the less-developed countries whenever human rights are equated with human numbers. Efforts to curtail immigration from poor overpopulated countries have aroused controversy in the United States and Western Europe. Americans urge the principle of "one man/one vote" upon the rest of the world. How would they respond to a demand for massive immigration from crowded nations on the principle of "one person/one acre?" The effects of developments in biotechnology upon this sensitive area of population policy are not easy to foresee. But if people are free to follow their personal preferences with the aid of widely available and reliable contraceptive technology and if international educational programmes relating to health, maternal and child welfare, reproduction technology, and birth control are effective, individual humans may eventually be able to make rational decisions regarding reproduction which express their interests and not necessarily those of their government or religious leaders.

Agriculture and Food

Impacts of biotechnology upon the reproduction of plants have been significant, reacting back upon animal life which ultimately depends upon the photosynthesizers that ultimately in some form feed most of the living world. With plant life the biotechnologists have freer hands than with humans and many other animals. If developed with care to preserve and protect indigenous plant species and their native ecosystems, plant genetics and biotechnology offer possibly the best prospect for many tropical countries to improve the health and nutrition of their populations.[17] Industrial applications involving biomass, as in ethanol fuel production, are promising, especially in countries with abundant rainfall and sunlight. Biotechnology offers tangible hope of raising the quality of life in those countries which have the necessary ecological conditions but lack the resources needed for industrial development based on oil, coal, or atomic energy. Capital costs of biotechnology could be relatively low, but a requirement for its development is investment in education in the life sciences.

Agriculture, including the closely allied disciplines of horticulture and silviculture, has been the principal field of modern plant biotechnology. Microbial biotechnology is as old as agriculture—as old as the discovery of the art of making alcoholic drinks through the fermentation of fruits and grains. Again, as has been emphasized, the interactive character of living material has enabled microbiology to assist medicine through the processing of bacterial and fungal life forms, as with penicillin, in a growing list of biologically derived compounds.

Agriculture was the object of the first large-scale scientific application of biotechnology.[18] Private and governmental efforts to improve domestic plant and animal stocks have been carried on for centuries. A large advance was made when in 1862 the Congress of the United States through passage of the first Morrill Act initiated the system of agricultural colleges and experiment stations that brought the United States to a position of leadership in the production of food and fiber.[19] American agricultural experts have brought their knowledge and skill to many countries of the world, through both governmental (e.g., U.S. AID) and private (e.g., Rockefeller Foundation) sponsorship.

The United Nations Food and Agriculture Organization (FAO), established in 1949, provides a global structure for communication, cooperation, and application in various aspects of applied biology and biotechnology. The broad scope of FAO includes marine life, and it provides the coordinating machinery for the various regional fisheries commissions which make decisions regarding techniques of fishing as well as restrictions on catch. Forest practice and technology are also assisted by FAO which has sponsored world forestry conferences and encouraged regional forestry agreements, for example, in equatorial Africa.[20] Control of locusts is another example of international cooperation in which FAO has been involved. The migration of vast numbers of locusts across large areas of Africa and Western Asia necessitate participation of all affected states in control measures if any are to be spared the ravages of locust invasions.[21] In time, biotechnology may provide ways to control this problem.

The concerns of FAO and the regional fisheries commissions with life cycles and populations of marine animals are more specifically addressed by the responsibilities of the International Whaling Commission for conserving the stocks of cetaceans. In these and other aspects of wildlife management, a variety of agreed-upon life science techniques for estimating population levels, determining breeding and survival patterns, and studying migratory behavior are essential to international cooperation.[22]

On 7 April 1982 a Convention for the Protection of Antarctic Marine Living Resources came into force, but its effectiveness depends heavily upon biological research and surveillance. To accomplish this purpose, the project BIOMASS (Biological Investigations of Marine and Antarctic Systems and Stocks) has been established. It is an international cooperative research program sponsored by two committees of the International Council of Scientific Unions and by the Intergovernmental Oceanographic Commission (associated with UNESCO).[23]

Bioscience and biotechnology are incorporated in measures to protect plants and animals from imported contagion. Plant and animal quarantines are legal provisions, but the tests and immunizations associated with international trade involve applied biology. Phytosanitary certification of plant materials in international commerce may in a broad sense be regarded as an application of biological science. International agreements such as the International Plant Protection Convention signed at Rome on 6 December 1951, the Plant Protection Agreement for

Southeast Asia and the Pacific region signed also in Rome (27 February 1956), the Soviet bloc Agreement Concerning Cooperation in the Quarantine of Plants and then Protection against Pests and Diseases (14 December 1959), and Plant Protection Commissions in the Near East and Caribbean regions depend in part for their effectiveness upon biotechnology in relation to plant genetics and plant pathology.[24]

Biological Variety and Diversity

The preparedness of national governments to cooperate in international environmental and ecological programs depends heavily upon agreed criteria for ascertaining and interpreting facts. There appears to be a reciprocal relationship here between reliable bioscience technologies and international agreements. Where national governments must act against their perceived national interests, or the wishes of politically influential constituents, as in the setting of catch limits for fish or quotas for the taking of whales or fur seals, it is important that cooperative action be defensible by scientific tests and measurements.

Nowhere is this more apparent than in enforcement of the Convention on International Trade in Endangered Species of Wild Fauna and Flora (CITES) in force since 1975. More than basic biology is required to ascertain the status of various species and in this effort, as in others already identified, policy needs stimulate the search for the development of new biology-related technologies. Techniques from simple banding of migratory birds to sophisticated telemetry applied to many wide-ranging species are, at least in part, scientific responses to the need for guidance in policy making.

The interactive effect of biotechnologies may be illustrated by the efforts of scientists, governments, and nongovernmental international organizations to protect and preserve ecological variety and diversity. In modern times, and especially in recent decades, human invasion and alteration of natural ecosystems have been major factors in loss of plant and animal species. Concern over species loss has been both national, and international, but increasingly the latter as international commerce, international economic development, and human migration and resettlement schemes have changed the face of the earth. National and international policies to prevent or retard ecological impoverishment depend upon valid knowledge of what is actually happening. In 1963 the International Union of Biological Sciences and the International Council of Scientific Unions lauched a ten-year International Biological Programme (IBP) to study ". . . organic production on the land, in fresh waters, and in the seas, so that adequate estimates may be made of the potential yield of new as well as existing natural resources, and human adaptability to changing conditions."[25] The objective was to inform government planners in the hope that action might follow information.

IBP studies enlarged the foundation for subsequent international efforts of which the 1968 Biosphere Conference held in Paris under the sponsorship of the United Nations Educational, Scientific, and Cultural Organization (UNESCO) and associated international organizations was perhaps the most important. The agenda and final report of the Biosphere Conference broadened the range of biological concern beyond the scope of the IBP. Among other results, it was followed by establishment in association with UNESCO of Man and the Biosphere Programme (MAB).[26]

Possibly the most significant accomplishment of MAB has been the establishment of biosphere reserves in cooperating countries under standards developed through the MAB Directorate on the Conservation of Natural Areas and Genetic Material and International Coordinating Council elected by the UNESCO General Conference. Many protected areas and ecological reserves have been established apart from MAB, yet as of 1986, it accounted for 252 official biosphere reserves, designated in 65 different countries.[27]

In recent years a new threat to biological diversity and variety appears to have arisen from advances in biotechnology itself. Experimental genetics and plant and animal breeding have produced new types of living species adapted for various human purposes. Manipulated plant and animal germ plasm are now widely distributed through international commerce. Fears have been expressed that these new strains, particularly of patented seeds for grain production, will displace naturally evolved genetic material to the point of possible extinction.[28] As nations move from subsistence or barter agriculture to cash crop monoculture for international markets using "green revolution" and "agribusiness" techniques, many potentially valuable natural strains of plant species may disappear. To forestall this consequence efforts have been made to establish artificial (laboratory) germ banks to preserve genetic material. Biosphere or natural area reserves have been regarded by some biologists as a more dependable means of genetic preservation. Forest reserves would seem highly important for many reasons, not least of which is the potential impact of genetic technology on silviculture.[29]

In response to the issues raised by genetic innovation and threats to biological diversity and variety, several international and national initiatives have been undertaken in addition to those associated with MAB. The importance of agriculture in international development programs led to the establishment of a Consultative Group for International Agricultural Research (CGIAR) associated with the World Bank, FAO, and the United Nations Development Programme (UNDP). In 1974 CGIAR, FAO, and the United Nations Environment Programme (UNEP) established the International Board for Plant Genetic Resources (IPBGR), not, however, a U.N. agency. IPBGN has emphasized genetic preservation through seed banks, and a similar board for animal genetic resources has been urged.[30] Thus two approaches to genetic conservation are thus being employed, both requiring international cooperation—biological reserves and genetic banks.

The commercialization of biotechnology is thus extending or enlarging the dimensions of international biopolicy. For example, in 1961 representatives of twelve of the more technologically advanced countries established the International Union for the Protection of New Varieties of Plants (UPOV).[31] The purpose of UPOV is to protect private (or national) rights to patented plant material. But this effort has been opposed in the less-developed countries which tend to regard genetic resources, even though manipulated through biotechnology, as belonging to "the common heritage of mankind." In November 1981 at the Twenty-First Conference of FAO, Mexico proposed an international convention, not only to ensure the conservation of genetic resources for agriculture, but also to guarantee their free availability and exchange among all nations. This move was successfully countered by the more-advanced nations, and a nonbinding resolution was adopted instead. The equities involved are confused and in dispute.

The implications of the variety-diversity issue for bilateral international technical assistance were addressed in the U.S. Strategy Conference on Biological Diversity held in Washington, D.C., 16–18 December 1981 under the sponsorship of eight federal agencies and the United States Man and the Biosphere Program.[32] The issue has also received a high place on the agendas of nongovernmental organizations, for example, in the United States in the Global Tomorrow Coalition of conservation and environmental organizations and internationally in the World Conservation Strategy of IUCN and UNEP. The United Nations General Assembly also addressed the issue in the World Charter for Nature adopted on 28 October 1982 declaring that:

> The genetic viability on earth shall not be compromised; the population level of all forms, wild and domesticated, must be at least sufficient for their survival, and to this end necessary habitats shall be safeguarded.[33]

The imperatives of the charter cannot easily be translated into national policies. The significance of the charter is rather in its statement of a pro forma consensus among nations regarding obligations toward the biosphere. Even the nations that voted against adoption of the charter (the United States) or those that abstained (the eighteen signatories to the Amazonian Treaty) did not oppose its basic principles or intent.

The practical effect of the charter may be small, yet it symbolizes the arrival of the governments of the world at a point of consensus in principle that has been previously approached only in the Declaration and Principles adopted in 1972 by the United Nations Conference on the Human Environment.

Summation

Biotechnical events considered especially newsworthy, such as organ transplants, artificial hearts, or test-tube babies, are widely publicized—but with little consideration given to their social implications. The implications of biotechnology for agriculture receive much less attention, although everyone everywhere might ultimately be affected by its consequences. Least understood or appreciated are the interactive effects of biotechnologies on international relationships. Knowledge concerning these effects depends upon a degree of interdisciplinary collaboration which is exceptional and largely unrewarded in modern science.

Scientific knowledge and technology have never been wholly or permanently confined (or barricaded) by national frontiers. Even so, the realm of the noosphere—the continuing realm of concepts and ideas—is not uniformly distributed around the globe. There are intellectual deserts and formidable obstructions hampering the free flow of knowledge. Yet especially when applied in forms perceived as useful, knowledge spreads—penetrating any society in which it is culturally receivable. Biology-related technologies vary widely in acceptability. Some have been avidly sought, notably preventatives for infectious disease; others such as birth control technologies have been banned by some governments, but have become widely disseminated nonetheless with increased worldwide communications.

To believe that the international flow of biotechnology is free from political manipulation, commercial self-interest, ethnic suspicion, and religious opposition would be naive. The issues sampled here, and many others that might have been reported, have often led to international tension—sometimes to contention. The variety and diversity of the issues preclude any simple governing policy or agreement among nations. The issues can be dealt with separately as they arise, but principles or guidelines upon which particular issues may be considered and evaluated may be amenable to general agreement.

Some years ago, persuasive argument might have been made that principles and standards must await prior formulation at national levels. Today that argument is less persuasive for the following reasons. First, where international policies are necessary or desirable, international agreements may more easily be obtained if nations attack the policy issue jointly at a point at which concern is general; the ability of nations to cope independently with the issues varies greatly. For many nations the only effective national policy for the rational control of biotechnology would be through international collaboration.

A second reason for early attention to international agreement upon principles is the rapidity with which biotechnology may be expected to expand. Humans cannot safely wait for national politics in each country to sort out the relevant issues and identify fundamental principles of common concern. Novel situations and issues continue to arise, and the possibility of serious and complicating international developments cannot safely be ruled out. Biotechnology has tended to be an enemy of conventional assumptions.

No nation has arrived at a level of bioscientific sophistication that clearly makes it an exemplar to others. Some nations are advanced (over others) in the development of bioscientific technologies, but not all governments of scientifically advanced states are receptive to specific applications or to international cooperation. An advantage of international deliberation upon developments of mutual concern is that there may at present be less occasion for fear of cultural or commercial domination when nations are collaborating in common efforts to establish effective policies for the beneficial uses of biotechnology. The account of the development of the United Nations Fund for Population by Rafael M. Sales subtitled "The Multilateral Approach to Population" (1976) suggests that much of the success of international cooperation in the highly sensitive area of population policy was the international character of the effort from its very inception.[34]

National issues today may be internationalized tomorrow, for example, under principles of human rights. Life science knowledge and technology are already destabilizing conventional human rights and criminal justice assumptions. Abuses of biotechnology by both dictatorial and totalitarian regimes may cause some persons to turn away from any consideration of linkages between politics, public policies, and biotechnology. But unbiased scientific inquiry and innovation have not been responsible for past abuses, and indeed the absence of an international forum for nonpolitical consideration of problematic issues may have been a factor allowing abuses to occur. A persuasive reason for early international action is the double-edged character of many of the issues involved—great potentialities for good but perhaps entailing risks of inhumane or ecologically disastrous consequences.

An international institution adaptable to an inquiry into the international implications of biotechnology is needed, and one may be developing. At least two efforts to establish an international system for the development and dissemination of biotechnology have been under way. Meeting in Madrid in mid-September 1983 under sponsorship of the United Nations Industrial Development Organization (UNIDO), ministerial-level delegates from twenty-five nations adopted a set of statutes for an International Centre for Genetic Engineering and Biotechnology (ICGEB).[35] Planning for the center was undertaken primarily by UNIDO pursuant to its mission to aid industrial development in the less-developed countries. In recognition that applications of genetic engineering and biotechnology in vital fields such as health, energy, or food are of major relevance to developing countries, an ultimate objective of the center is to redress the imbalance in scientific capability between the richer and poorer nations. In October 1986 the center entered an operational phase with agreements signed in Vienna providing for dual headquarters in New Delhi, India and Trieste, Italy.[36] An important function of the new center is *bioinformatics*—development of a data bank and communications system of particular value to less-developed countries.

Support for ICGEB has come almost wholly from less-developed nations. The United States, Japan, and the larger West European states have not shown great interest. Objections to the center and to intergovernmental involvement generally include (1) doubts that the center would produce first class scientific work, especially if located in a Third World country and (2)—largely American—that any intergovernmental collaboration in biotechnological research would interfere with market forces. Meanwhile, however, an International Technology Network has been proposed by a group of scientifically advanced countries led by France and the United Kingdom with the participation of Canada, Italy, and Japan, with the possibility of full participation by the European Community, the United States, and the German Federal Republic.[37]

International collaborative and cooperative arrangements have been growing rapidly among the developed countries. In many of these relationships, governments and nongovernmental organizations are jointly concerned. The European Community has established a Biotechnology Regulations Interservice Committee which began work in September 1985 to review regulations on commercial applications of biotechnology, to identify existing laws and regulations that may be applied to commercial biotechnology, to review the guidelines for RDNA research, to clarify the regulatory paths for biotechnology products, and to determine if current regulations are sufficient to deal with possible risks, and initiate specific regulatory actions where necessary. The Organization for Economic Cooperation and Development has developed guidelines for the safe release of genetically altered microorganisms by its members, but not without objections from the United States. The number of bilateral and multilateral international agreements is large, growing, and includes government to government, government to private organizations, and private to private firms and scientific institutions. Nations primarily involved are the United States, Great Britain, Canada, France, Switzerland, West Germany, Netherlands, and Japan.[38]

Meanwhile lesser developed countries have been negotiating their own biotechnology cooperation. For example, in Latin America, Interciencia has created a continuing committee for networking in biotechnology with particular reference to an inventory of biotechnology researchers, publications, workshops, and establishing national and regional priorities.[39] There are many reasons for developing countries to acquire competence in the field of biotechnology and there are indications that many of them will do so.[40]

In summary, biotechnology with its potential for changing, perhaps irreversibly, living systems and species, is today largely concentrated in a few countries and within certain organizations or institutions within those countries, notably in universities, research laboratories, experiment stations, and in corporate firms developing commercial applications. Innovations developed in these institutions may affect the entire world. Yet nations and their peoples are only beginning to become aware of the possibilities inherent in the rapidly expanding fields of the life sciences. But biotechnological innovation has now attained a potential for benefits but also for unwanted effects that cannot be managed wisely or effectively in the absence of some system of international responsibility. How to discover and develop such a system is a challenge to political innovation that will not be easily met. The consequences of failure to find some concerted means for rational international deliberation on biopolitical issues is likely to be the loss of opportunities for realization of benefits and the increased likelihood that biopolitical invention may be diverted to destructive uses.

Advances in biology upon human understanding, attitudes, and values may be as great, perhaps greater, than the changes induced by biotechnology itself. Biological information and its influence upon concepts and mores are not separable from biotechnologies, but they are different. Two of the great technoscientific advances of our time—biotechnology and information technology—have powerful mutually reinforcing capabilities that may induce changes in societies with a scope and speed without precedent. How people perceive and understand life almost certainly influences how they will act in relation to biotechnology. The very meaning of what constitutes a human person and of the right of humans to procreate are today being affected by cumulating findings in the life sciences. Knowledge for which humans have no precedent may lead to social and political developments for which no historical analogies exist. We cannot now foresee how advances in the life sciences may ultimately affect relations among nations, but it would be short-sighted to believe that the effects will be insignificant.

Notes

1. For purposes of social and political analysis, this loose definition seems preferable to the more precise one proposed by the authors of an OECD monograph: "the application of scientific and engineering principles to the processing of materials by biological agents to provide goods and services" (Alan T. Bull, Geoffrey Holt, and Malcolm D. Lilly, *Biotechnology: International Trends and Perspectives* [Paris: Organization for Economic Cooperation and Development, 1982], 21). For other definitions and sources see chapter 8, note 1.

2. See F. E. Zeuner, *A History of the Domestication of Animals* (London: Hutchinson, 1963). Pharmaceutical technologies have been used to tranquilize animals, and experimental work has been undertaken to cope with the problem of excess reproduction of wild herds (e.g., wild horses) through injections to counteract fertility.

3. Kenneth Dahlberg, *Beyond the Green Revolution: The Ecology and Politics of Global Agricultural Development* (New York: Plenum Press, 1979). See also Jack Doyle, *Altered Harvest: Agriculture, Genetics, and the Fate of the World's Food Supply* (New York: Viking, 1985).

4. The Iranian poultry incident was reported to the author by U.S. AID workers during a visit to Teheran. The report is plausible and other examples of resistance to innovation are not hard to find. E.g., Anacleto Apodaca, "Corn and Custom: The Introduction of Hybrid Corn to Spanish American Farmers in New Mexico," in Edward H. Spicer, ed., *Human Problems in Technological Change: A Casebook* (New York: Russell Sage Foundation, 1952), 33–39

5. E.g., see William B. Hewitt, *Plant Health and Transfer of Genetic Resources* (Cleveland, Ohio: CRC Press, 1977); and *FAO Plant Protection Bulletin* for reports on phytosanitary regulations and legislation. For the International Plant Protection Convention of 1951, see *U.N. Treaty Series* 149, no. 1963 (1952): 68. For regional agreements and treaties, see *U.N. Treaty Series* 247, no. 1963 (1956): 400, and 422, no. 6067 (1962): 33, 42. National protective legislation has also been adopted in most countries; e.g., in the United States the various statutes and regulations administered by the Animal and Plant Health Inspection Service of the Department of Agriculture.

6. See Gerry J. Elman, "Environmentalists Ask Court to Halt Testing of Recombinant Plant Symbiotes in the Field," *Genetic Engineering News* 3 (November–December 1983): 1; and Keith Schneider, "Biotech's Stalled Revolution," *The New York Times Magazine*, 16 November 1986, 42 ff. See also "Moratorium on Approval of Deliberate Release Experiments Requested," *Chemical Work* 136 (May 1986): 48.

7. Keith Schneider, "Gene-Splicing Tested in Argentina," *San Jose Mercury News*, 11 November 1986, 3D. See also Stephen Budiansky, "Army Pall over Argentine Science," *Nature* 311 (20 September 1984): 201 ff.

8. David Pruett, *Human Fertility, Health and Food: Impact of Molecular Biology and Biotechnology* (New York: United Nations Fund for Population Studies, 1984). On biotechnological applications to health in developing countries, see Marc Lappé, "Biotechnology's Debt to Public Health," *Technology Review* 88 (August–September 1985): 14 ff.; Stanford University News Release, 10 December 1984; David Marsh, "China and Taiwan Sign Biotechnology Pacts with France," *Financial Times of London,* 5 April 1985, 4. On environmental effects on health, see William D. McKee, *Environmental Problems in Medicine* (Springfield, Illinois: Charles C. Thomas, 1974). For cases in point, see "Lassa Fever," NOVA #1003, Boston: WGBH Educational Foundation, transcript; "The Cancer Detectives of Lin Xian," NOVA, Boston, WGBH Educational Foundation, 1983, transcript; Peter Clifford, "Carcinogens in the Nose and Throat: Nasopharyngeal Carcinoma in Kenya," *Proceedings of the Royal Society of Medicine* 65 (August 1972): 682–85; and more generally, John Lenihan and William W. Fletcher, *Health and the Environment* (New York: Academic Press, 1976); Erik P. Eckholm, *The Picture of Health: Environmental Sources of Disease* (New York: W. W. Norton, 1977); articles in the *Archives of Environmental Health* 1960–present; and publications of the World Health Organization.

9. The following references define issues and present evidence but are not to be regarded as thorough assessments: Robert Richter, "Pesticides and Pills: For Export Only," Parts 1 and 2, 7 October 1981, transcript of television broadcast on Public Broadcasting Service (U.S.); Ruth Norris, A. Karim Ahmed, S. Jacob Scherr, and Robert Richter, eds., *Pills, Pesticides, and Profits: The International Trade in Toxic Substances* (Croton-on-Hudson, New York: North River Press, 1982); David Weir and Mark Shapiro, *Circle of Poison: Pesticides and People in a Hungry World* (San Francisco, California: Institute for Food and Development Policy, 1981); and Gaston Vettorazzi, *International Regulatory Aspects for Pesticide Chemicals* (Boca Raton, Florida: CRC Press, 1979). In addition to the foregoing, see Barry Castleman, "The Export of Hazardous Factories to Developing Countries," *International Journal of Health Services* 9, no. 4 (1979): 569–606; and House Committee on Foreign Affairs, Subcommittee on International Economic Policy and Trade, *Hazardous Product Exports,* 97th Congress, 1st session, 12 March 1981.

10. "International Code of Marketing of Breast Milk Substitute," *WHO Chronicle* 35 (1981): 133–34; *Keeslings Contemporary Archives,* 21 August 1981, 31044.

11. U.N. General Assembly, "Report of the Economic and Social Council, Protection against Products Harmful to Health and Environment," A/C.2/37/L65/Rev.1, 7 December 1982; and Libby Bassett, "U.N. Votes To Protect People from Harmful Imports," *World Environment Report* 9 (15 January 1983): 3.

12. See *Federal Register* 46 (19 January 1981): 4659–64.

13. "The Alma-Ata Conference on Primary Health Care, " *WHO Chronicle* 32, no. 11 (November 1978): 428; and *Measuring Change in Nutritional Status: Guidelines for Assessing the Nutritional Impact of Supplementary Feeding Programs for Vulnerable Groups* (Geneva: World Health Organization, 1983).

14. See Robert H. Blank, "Genetic Screening," *The Political Implications of Human Genetic Technology* (Boulder, Colorado: Westview Press, 1981), 46–62, includes references.

15. Rafael M. Salas, *People: An International Choice, The Multilateral Approach to Population* (New York: Pergamon Press, 1976), and more recently *Reflections on Population*, 2d ed. (New York: Pergamon, 1985).

16. United Nations, *Report of the United Nations World Population Conference*, 1974, held at Bucharest, 19–20 August 1974 (New York: United Nations, 1975), 38, 43, 44, and *Report of the International Conferencce on Population*, 1984, held at Mexico City, 6–14 August 1984 (New York: United Nations Department of Technical Cooperations for Development, 1984).

17. See E. J. Da Silva, "The Renaissance of Biotechnology: Man, Microbe, Biomass and Industry," *Acta Biotechnologia* 1 (1981): 207–46; and D. O. Hall and E. J. Da Silva, "Photosynthesis: A Bio-Solar Tool for Development," *Nature and Resources* 19 (April–June 1983): 2–9.

18. Charles B. Heiser, *Seed to Civilization: The Story of Man's Food* (San Francisco, California: Freeman, 1973).

19. See Edward Wiest, "The Land-Grant Colleges," chapter 10 in *Agricultural Organization in the United States* (Lexington: University of Kentucky Press, 1934, reprinted by Arno Press, 1975); and John M. Gaus, Leon O. Wolcott, with a chapter by Verne B. Lewis, *Public Administration and the United States Department of Agriculture* (Chicago: Public Administration Service, 1940), especially 5–7 and 368–74.

20. E.g., see reports on Eighth Forestry Congress (1978) in *Unasylva* 30, no. 121 (1978): 35–38, and *Uniterra* 3 (November 1978): 1–2. Also, "African States Agree on Forests Convention," *Uniterra* 5 (July 1980): 2.

21. B. P. Uvarov, "Efforts to Control Locusts in Africa Described," *Science* 130 (4 December 1959): 1564–65; and Stanley Baron, "No Frontier in the Fight against the Desert Locust," *Ceres* 1 (September–October 1968): 32–42.

22. Extensive references on national and international policy for marine life are provided in Lynton Keith Caldwell, *International Environmental Policy* (Durham, North Carolina: Duke University Press, 1984). See specifically Bryan J. Rothschild, ed., *World Fisheries Policy: Multidisciplinary Views* (Seattle: University of Washington Press, 1972); and *Mammals in the Sea*, 4 vols. (Rome: U.N. Food and Agriculture Organization, 1981).

23. On Biological Investigations of Marine and Antarctic Systems and Stocks (BIOMASS), see F. M. Auburn, "Krill," chapter 7, and indexed references in *Antarctic Law and Politics* (Bloomington: Indiana University Press, 1982), 205–67. Also, Sayed Z. El-Sayed, "SCAR/SCOR Conference on Living Resources of the Southern Ocean," *Antarctic Journal* 12, nos. 1–2 (1977): 1–4; and "The Southern Ocean," *Oceanus* 18 (Summer 1975): entire issue, especially Gerald S. Schatz, "A Sea of Sensitivities," 40–55.

24. See note 5 above.

25. See "International Biological Programme: Report of the Planning Committee," 15 November 1963 *BioScience* 14 (April 1964): 43–49; and E. Barton Worthington, ed., *The Evolution of IBP: The International Biological Programme* (Cambridge: Cambridge University Press, 1975), and Worthington, "The Twentieth Anniversary of IBP," *Environmental Conservation* 9 (Spring 1982): 70.

26. *Intergovernmental Conference of Experts on the Scientific Basis of Rational Use and Conservation of the Resources of the Biosphere*, SC/MD/9, final report

of conference held at UNESCO House, 4–13 September 1968 (Paris: UNESCO, January 1969); and the international cooperative program of Man and the Biosphere (MAB) which followed. On the "Biosphere Conference," in retrospect, see Francois Bourliere and Michel Batisse, "Ten Years after the Biosphere Conference: From Concept to Action," *Nature and Resources* 14, no. 3 (1978): 14–17. On MAB, see Batisse, "The Relevance of MAB," *Environmental Conservation* 7 (Winter 1980): 179–84, and "The Future of MAB," *Environmental Conservation* 9 (Spring 1982): 71–72; also, Frank B. Golley, "Ten Years of MAB: Establishing a Balance Sheet," *Nature and Resources* 17 (April–June 1981): 3–7. A declared purpose of the U.S. National Environmental Policy Act of 1969 is to "promote efforts which will promote or eliminate damage to the environment and the biosphere . . . " (42 U.S.C., 4321–47, 1 January 1970).

27. United Nations Educational, Scientific and Cultural Organization, *Implementation of the Action Plan for the Biosphere Reserves*, 121 Ex/10 (Paris: UNESCO, 9 April 1985). See also *Nature and Resources* 20 (October–December 1984): 11–22; *infoMAB*, no. 7, 1987, 12. See Batisse, "The Biosphere Reserve: A Tool for Environmental Conservation and Management," *Environmental Conservation* 9 (Summer 1982): 101–11, and "Developing and Focusing the Biosphere Reserve Concept," *Nature and Resources* 22 (July–September 1986).

28. See O. H. Frankel and E. Bennett, eds., *Genetic Resources in Plants: Their Exploration and Conservation* (Oxford: Blackwell Scientific Publications, 1970). See also J. T. Williams, "Genetic Conservation of Wild Plants," and Robert and Christine Prescott-Allen, "The Care of or In Situ Conservation of Crop Genetic Resources," *Nature and Resources* 18 (January–March 1982): 14–20; John Walsh, "Germplasm Resources Are Losing Ground," *Science* 214 (23 October 1981): 421–23; Kenneth A. Dahlberg, "Plant Germplasm Conservation: Emerging Problems and Issues," *Mazingira* 7, no. 1 (1983): 14–25; D. L. Plunkett, et al., "Crop Germplasm Conservation and Developing Countries," *Science* 220 (8 April 1983): 163–169; J. Trevere MacFadyen, "A Battle over Seeds," *Atlantic Monthly* 256 (November 1985): 36–44. For a "popular" assessment, see Mark Shapiro, "Seeds of Disaster," *Mother Jones*, 12 December 1983, 12 ff. Note also FAO Conference Resolution 6/81, "Plant Genetic Resources," *Report of the Conference of FAO*, Twenty-First Session, Rome, 7–25 November 1981 (Rome: FAO, 1981), 27–28.

29. Gene Namkoong, "Genetics and the Forests of the Future," *Unasylva* 38, no. 2 (1986): 2–18.

30. On CGIAR, see *Europa Yearbook: 1983* 1 (1983): 81. On the International Board for Plant and Genetic Resources, see *Yearbook of International Organizations* (1983/84): E3150. See also R. H. Kemp, "Exploration, Utilization, and Conservation of Genetic Resources," *Unasylva* 30, nos. 119–20 (1978): 10–16, extensive references.

31. On UPOV, see *Yearbook of International Organizations*, 23d ed., (1986–1987): FF2671g. Note Sidney B. Williams, Jr., "Protection of Plant Varieties and Parts as Intellectual Property," *Science* 225 (6 July 1984): 18–23. See also "Animals Can Be Patented," *Science* 236 (10 April 1987): 144.

32. *Proceedings of the U.S. Strategy Conference on Biological Diversity*, 16–18 November 1981 (Washington, D.C.: Department of State, Publication 9262, April

1982). See also Gabor Vida, "Genetic Diversity and the Environmental Future," *Environmental Conservation* 5 (Summer 1978): 127–32.

33. U.N. General Assembly, Resolution Adopted by the General Assembly, World Charter for Nature, A/Res.37/7, 9 November 1982. Text reprinted in *Environmental Law and Policy* 10 (January 1983): 31–32. See also Libby Bassett, "UN Votes to Adopt World Nature Charter," *World Environment Report* 8 (30 November 1982): 2–3.

34. See note 15 above.

35. David Dickson, "UNIDO Hopes for Biotechnology Center," *Science* 221 (30 September 1983): 1351–53. See following publications of the United Nations Industrial Development Organization (UNIDO): *The Establishment of an International Center for Genetic Engineering and Biotechnology (ICGEB): Report of a Group of Experts,* UNIDO/15.254, distribution limited; *Elements of Some National Policies for Bio-technology: Note by the Secretariat of UNIDO,* UNIDO/15.270/ Rev.1., distribution limited.

36. See "Three Giant Steps for International Gen-Tech Center: $18M Interim Program Finalized," UNIDO News Release, (New York: UNIDO, 22 October 1986); Dermot A. O'Sullivan, "Focus of International Biotech Center Shifts from Politics to Science," *Chemical and Engineering News* 63 (15 April 1985): 23; Burke K. Zimmerman, "India and Italy Force Compromise on International Biotech Center Site," *Genetic Engineering News* 4 (May–June 1984): 3; Carl-Goran Heden, "Bioinformatics," *United Nations Industrial Development Report,* ID/WG 382/2/ AOD.6, 20 September 1982; also by Carl-Goran Heden, "New Mechanisms for Research Cooperation in Biotechnology," *Bio/Technology* 1 (November 1983): 770.

37. "Biotechnology Network Planned," *Science* 221 (30 September 1983): 1352.

38. See Robert Walgate, "Community Seeks New Funds," *Nature* 311 (18 October 1984): 594; "The European Commission's Approach to Biotechnology Regulation: European Biotechnology Information Project," EBIP News Release, January 1986; "U.S. Panel Split on Guidelines," *Nature* 318 (19 December 1985): 594; "Consider Biotech's Foreign Policy Potential," editorial, *Bio/Technology,* editorial 1 (October 1983). See also "Biotechnology: Ambition and Track Record," *Nature* 309 (7–13 June 1984): 500; *Biobusiness World Data Base* (New York: McGraw-Hill, 1983).

39. See Christian Orrego, "New Cooperative Effort Planned for Latin American Biotechnology," Costa Rica Meeting Report *Bio/Technology* 1 (July 1983): 413; and "Interciencia Begins Network of Biotechnology in the Americas," *Science* 224 (6 April 1984): 49. On Interciencia see *Yearbook of International Organizations,* 23d ed., 1 (1986–1987): DD4312.

40. See Sanford Lackoff, "Biotechnology and the Developing Countries," *Politics and The Life Sciences* 2 (February 1984): 151–83; and H. Monte Hill, "Biotechnology: The Key to Third World Development," a critique of Lackoff's article, *Politics and the Life Sciences* 3 (August 1984): 84–86.

10
Biocrats and Democrats: Allies or Adversaries?

No society will maintain itself indefinitely if the premises on which it proceeds are invalid. It may survive vicissitudes and hold together for long periods of time if the challenges it encounters do not exceed its response capabilities. Failures of societies to meet challenges of events and environment have been conjectured by archaelogists and examined by philosophers of history; most extensively by Arnold Toynbee.[1] If for whatever reason the circumstances within which the society has developed are altered beyond the society's adaptive responses, its disintegration, displacement, or transformation follows. This is not to say that all failures of society are caused by adherence to false premises, nor that it is always possible to discover what governing premises were false, nor what combination of circumstances may have caused previously harmless beliefs to become harmful. Miscalculations of leaders and governments are documented by historical evidence, but errors of judgment have been frequently made within a context of invalid assumptions which gave the judgment plausibility, even contemporary acceptance as "wisdom."

Historical societies have often succumbed to military conquest, but defeat has as often occurred after inward decline in shared values and cohesiveness, and misinterpretation of environmental circumstances. Economic, ecological, and demographic factors may affect societal decline, although a society may continue to retain vitality for a period of time. If, however, the society is divided against itself or misguided by its unexamined assumptions, it becomes vulnerable to overthrow from within or takeover from without. The foregoing circumstances in combination almost inevitably lead to downfall (the Montezuma syndrome).[2] Both were present and fatal in the conquest of Mexico by an incredibly small band of Spaniards. Societies have, however, apparently disintegrated through their own behavior in the absence of external force. Environmental mismanagement leading to soil exhaustion, desertification, or endemic diseases such as malaria appear as probable factors in the collapse of Khmer, early Mayan, and Easter Island societies, among many others. Overpopulation, in relation to life-support capabilities, may also overstress societal resilience, opening the way to communal violence, impoverishment, and disease. Short of actual socioecological collapse; Bangladesh, Haiti, and Somalia may be examples.

Adverse present day circumstances could be ameliorated, although not wholly cured, through the capabilities that science and technology now make available. But social assumptions, values, and institutionalized commitments often limit what can or will be done. Many environmental effects of human behavior can be measured and sociopsychological effects are being identified and verified as, for example, in personal stress associated with adverse ambient and occupational conditions. Trends may be discovered and probable outcomes projected. Social premises and practices can be analyzed and their validity more often ascertained than at any previous period of time. And yet these capabilities are today rudimentary and incomplete. The definitive causal factors in the development and decline of societies, remain matters of uncertainty and dispute. How, for example, is the validity of a social premise to be ascertained when it embodies value postulates not amenable to empirical testing? And to what extent, if at all, especially in highly diverse societies, does an apparently dominant social order influence the character of an entire society? Why does a hitherto dominant group or class sometimes abdicate and surrender its position to a formerly subordinant population? How far can individualism, pluralism, ethnicity, and racial differences be carried without loss of social cohesion to a degree insufficient to hold the society together?

We are not well prepared to answer these questions, but they will never be answered if not first addressed. Possibly they can only be answered by the ordeal of history. Our conjectures of today may tomorrow be shown to have been mistaken, but right answers are seldom found without elimination of answers found to be wrong. Initial conjectures may also be right although their confirmation may be long delayed. The cumulative influences of the life sciences on humanity's future are conjectural, but because those effects could be very great, they are of legitimate concern for everyone who cares about the future. Some social scientists are concerned and have formed an Association for Politics and the Life Sciences, and the International Political Science Association has had panels on biology and politics.[3]

Biocracy and Ideology

Does the dynamic complexity of advanced human societies put them beyond the present limits of scientific understanding? Given the past growth of scientific knowledge and methodology, it would be presumptious to say that the limits cannot be extended. Yet there is widespread perception that our ability to apply science-based technology outpaces our understanding of its implications. The social and behavioral sciences which should logically address these implications are the least developed of the scientific disciplines—the least equipped to demonstrate cause-effect relationships. However, to contrast physical and social science as *hard* and *soft* is misleading. The metaphors relate to reliable statistical "facts" that make physical phenomena predictable to a degree not found in the infinitely more complex social sciences. More meaningful would be to contrast the relatively simple and complex sciences. The complexities of physics and chemistry are basic to the phenomena of the life sciences and ultimately have been shown to enter into circumstances accounting for social behavior. In the transition from the molecular and chemical level to biochemistry, neurology, psychopharmacology, and behavioral chemotherapy, very large leaps are taken toward infinite complexity. Ralph Gerard's hierarchy of complexity is a more meaningful way to understand relationships among the sciences than the misleading hard science-soft science dichotomy.[4]

A greatly complicating factor in understanding the behavioral aspects of human organization is culture. In time, advanced computer technology with input more comprehensive and verifiable than that available today may provide ways of identifying, disentangling, simulating, and forecasting probable consequences of social behavior. Forecasting should not, however, be regarded as "prediction." The counter-intuitive behavior of social systems is notorious and human creativity and also nature may hold hidden surprises.[5] Yet to foresee what *may* follow from a given course of action may nonetheless be advantageous. At present the factual basis for building social theory is unfirm, although not everywhere soft. Social science capability is growing, but, in addition to its inherent difficulties, its potential for social prognosis has been handicapped by self-imposed limitations. Many of the problems of building a more reliable social science are methodological, but the more intractable problems thus far appear to be psychological and ideological. Granting exceptions, social science may be said to have been held captive by social philosophy.

This is not surprising if one considers that social science has been rooted in moral philosophy. Modern social science has been "conditioned" by its own paradigm or accepted beliefs about what causes human societies to behave as they do. There are disputes within the social sciences over issues of theory and method, but a dominant orientation is evident and assumes the character of tacit moral conviction. Throughout history politics, religion, and moral philosophy have been closely linked, although sometimes in paradoxical relationships. When political forces have attacked religions or separated themselves from religious practice, they have usually done so out of commitment to another "faith" which may not have been seen as a religion in the conventional sense, but which fulfilled many of its functions nonetheless. In some respects modern social science adheres to an academic "religion." But whatever the viewpoints of social scientists (and biologists as well) their findings and beliefs are projected into a world of far more diverse belief patterns.

In the larger world today, Marxism, fascism, humanism, and liberal democracy have been regarded by the more passionate of their adherents with a faith that is quasi religious. As with the orthodox religions, these ideologies explain the relationship of the individual human being to society and to history. Each has its own prophets, codifiers, and martyrs; each is guided by ethical assumptions that are reflected in political action. They may be described collectively as civic or secular religions. Collectively they have one other common characteristic: they are belief systems relatively uninfluenced by the findings of science, although Marxism has been acclaimed as a "science of society." In the world of actual practice, however, no political theory or system is found in an ideal or "pure" form. One cannot easily discuss their practical effects upon human affairs without generalization, which, of course, inevitably falls short of complete truth.

There is a general tendency today to associate all political virtues with democracy and all vices with other forms of governance. In historical fact, democracies have taken many different forms. Any meaningful discussion of democracy must therefore specify particular relationships and procedures. When this is done, however, objections will invariably be made that the political order described is not democratic. For purpose of clarity the expression "liberal democracy" will be used in the following discussion to designate a political system open to almost unrestricted popular participation and with minimal restrictions on freedom of thought and action.

In relation to the sciences an important generalization distinguishes liberal democracy from the others. At least in theory it differs in its openness to change. More than the others it has been able to accommodate new findings regarding human nature and society without abandoning the chain of reasoning upon which its credibility depends. Although not based upon scientific evidence, liberal democracy in principle is not necessarily inconsistent with present science-based understandings and being less doctrinaire in content is less vulnerable to the advances in the sciences of human behavior. This does not mean, however, that at all times and places all biocratic ideas will receive social approval or acceptance. Democracy is not synonymous with broad-mindedness.

Biocracy is a term no less imprecise than those identifying the formal ideologies. But it is not a formal ideology—not yet, perhaps never—and is best described as the cumulative influence of the life sciences (and the life forces that they describe) upon society and public affairs. Nevertheless, a biocratic philosophy in fact, if not in name, is present and growing although not directly a significant political force. As a body of thought and theory about human nature and the status of man on earth, biocracy in effect challenges assumptions basic to all of the dominant ideologies, yet it offers no political theory and has nothing to say about forms of government. It threatens to undermine the premises of the ideologies, but thus far offers no prescriptions of its own to replace theirs.

The fundamental difference between biocracy and the ideologies is their placement of humanity in the cosmos. All of the now dominant political ideologies place mankind and human society at the center of things. They are homocentric or people-centered belief systems. Biocracy offers a biocentric or life-centered, man in and of nature, perspective on reality, a dichotomy analyzed by Paul W. Taylor.[6] At variance with the ideologies, biocracy does not separate humanity and its cultures from nature, from genes, DNA, and the biosphere. The nature-nurture or culture-environment dichotomies do not arise in biocentric thinking. They do not arise because they are understood as inextricably interrelated. Complex synergistic interrelatedness and evolutionary interaction characterize a biocratic view of the world which does not accept truths that are self-evident or man alone as the measure of all thing.[7]

Knowing little that was valid about biological influences on human behavior, historical ideologies found causes in external forces unrelated to internal influences of human mind and body. Human failures and successes have been largely explained by environmental factors external to the human individual. Marxism is possibly the most environmentalist of political ideologies, and it was not illogical for the environmentalist theories of Trofim Lysenko to have been adopted by Communist Party leader Joseph Stalin to the rejection of less accessible internalized genetic influences.[8] Nor was traditional Christianity unique among religions in attributing human error to external forces of evil that insinuated themselves into the human personality.

The premises of prevailing ideologies are essentially modern, with roots in the seventeenth and eighteenth centuries. Far older traditions accepted innate and reciprocal influences, that today might be regarded as genetic predispositions as explaining human behavior. Long before Claude Bernard, *les milieux intérieur et extérieur* were linked in ways that were often seen to be reciprocal. Traditional views of man and nature were very often holistic; internal and external aspects of life were recognized but not separated in the interpretation of human behavior. Human existence was a part of, not apart from, its environment. There was the fatalism of tragedy in ancient belief exemplified most vividly in Greek drama, that humans could not escape from the environment in which they were immersed and which, of course, included themselves. Science, notably life sciences, has enabled modern people to cope with interior-exterior interactions, some of which we describe as discomfort or disease, but it does not alter the fundamental situations of humans within the biosphere, or the need for humans to gain a more valid understanding of the external-internal forces that influence their behavior—individually and collectively.

This holistic concept of the whole man—biological and cultural interacting with environmental elements—has been basic to all of the essays in this book. It was fundamental to biologist Edmund W. Sinnott's essay on "The Biological Basis of Democracy" (1945). Sinnott believed that the true basis of human freedom lay in human diversity, but that evolutionary man was biologically conservative in the sense that the genetic endowment afforded the only mechanism by which this opportunity for freedom could in full measure be achieved. He wrote that:

> A modern generation, tough-minded in an age of disillusion, must be convinced that any plan for human society must have a sponsorship transcending the changing and fallible wisdom of mankind. It must conform to nature and her laws, must be in harmony with the very structure of the universe itself. In short, it must be grounded on scientific fact rather than on passion, emotion or intuition. For good or ill, science is now a great court of appeal.[9]

Rhetorical and Functional Democracy

Theoretical ideologies belong among the images in Plato's cave. They reflect certain realities that differ in many ways from their projections which may be idealized without fidelity to the reality that they purport to represent. Democracy, however, in nearly all its meanings, has the *possibility* of differing from the more coherent "closed" ideologies in the variety of its expressions, its openness to interpretation, and its tolerance for differing ideas. For example, socialism and monarchy may coexist within a democratic political system. Societies may also be democratic yet up to an uncertain point may exclude some significant portion of their members from political participation. How the excluded minority is governed and the nature of the rights that it possesses determines the extent to which the political order is democratic. Democracies may also establish dictatorships for limited periods under special circumstances without loss of fundamental democratic character. The exercise of presidential war powers in the government of the United States is a case in point.

Whether democracy and biocracy are compatible or antagonistic depends upon how each is defined in theory and realized in practice. Idealized and abstract descriptions of democracy give little insight into its various applications in reality. Biocracy may or may not be "democratic," depending partially upon how democracy is defined, and partially upon what implications for human society are inferred from findings in the life sciences. There is thus no single simple answer to whether biocrats and democrats are or might be allies or adversaries, and appellations are, of course, figures of speech that suggest tendencies rather than consistent points of view.[10]

In examining the compatibility of biocracy and democracy one must distinguish between rhetoric and practice. Rhetorical democracy is usually expressed in political oratory, polemics, and popular journalism, and tends to be long on adjectives and short on logic. It may find expression in lofty and impassioned prose as in the Declaration of Independence of the United States or the Rights of Man of revolutionary France. Pronouncements like these are intended to rally support and to justify action. They appeal to emotions and their recourse to human experience is highly purposeful and selective. Yet the practical expression of their sentiments may be far removed from any idealized literal expression. In historic fact human slavery coexisted with the Declaration of Independence and the Rights of Man proved no barrier to the Reign of Terror in the French Revolution.

In contrast to the tendencies of rhetorical democracy, functional democracy is essentially procedural and could be consistent with a range of some (but not necessarily all) biocratic propositions. Democracy can be defined with more meaning in functional rather than in theoretical terms. Democratic ideas and institutions can be tested and evaluated by direct reference to what they do, how they do it, and what then follows. Actions and their consequences may be examined if the society is sufficiently open to self-criticism. If the future of mankind is not deterministic, the course of history may be altered or reversed, and here the life sciences are beginning to develop understandings that could influence how social and political choices are evaluated, to enable people to take broader and longer views of their options. Trends may be revealed and consequences forecast without pretensions to prediction. Open democracies have, at least, the possibility of correcting their own mistakes and of evaluating alternative courses of action, free from ideological blinders.

Particular democracies may of course be closed to mass participation, may be adverse to social change, and may be intolerant of deviant opinion. Democracies need not be liberal to be in some sense "democratic." Like other forms of governance, democracies may be bounded and confined to a particular demos or polity. Thus, some sector of a society may be democratic in an otherwise undemocratic society. Classical democracies were characteristically small exclusive units but were capable of imposing undemocratic rules on people or communities outside their self-defined limits. The mass democracies of modern times tend to be open, pluralistic, and weakly committed to any ostensibly permanent structure or institutional expression. It is this looser generalized form of democracy that offers the greater prospect of adaptation to the influences of advances in the life sciences. But it also harbors the weakness of difficulty to achieve collective consensus or to achieve collective self-discipline and restraint.

The Biology of Behavior

Although biocracy may be compatible with liberal democracy, it is a potential adversary of every political order that is based on premises inconsistent with the forces that govern human behavior. No political theory has thus far been based—or could have been based—upon an adequate, verifiable understanding of human nature. All contemporary political theories have relied for their principles upon a priori reasoning and revelation, there being no science of human behavior sufficiently developed to be reliable. Since the advent of modernity, political parties have resisted the belief that human behavior could be explained by internal innate forces. External influences were identified as enlarging human capacities for achievement or accentuating human weaknesses. Hereditary traits, usually traits for the worse but sometimes for genius, were acknowledged, but their causes and mechanisms were not understood and their social significance largely denied.

Political behavior has thus far been explained largely as an effect of forces present in an externalized environment. Human nature has been tacitly accepted as a neutral entity to be shaped (with pathological exceptions in particular individuals) by environmental forces, natural and cultural. History records presumed influences of climate, terrain, institutions, and extraordinary personalities upon human societies, but it is a history limited by what was known and believed to be causal. Environmental explanations failed to explain when the environment was perceived as essentially external to an individual or a society and its internalized effects and responses to some extent acknowledged but not understood. To understand environment as indicative of a *relationship* is essential to any scientific or sophisticated explanation of its effects.

As we have noted, the French physiologist Claude Bernard described the interaction of the human interior and exterior environment as relational, a simple, observable but widely ignored relationship. Objects in themselves are neither environments nor environed, but are commonly misinterpreted by attribution to the properties of the objects themselves (e.g., climate, soils, microorganisms) with insufficient regard to the relationships among them. No man is an island and no islands are truly isolated from environmental relationships. That cause-effect conclusions have been drawn from the imperfect record of human experience does not mean that effects or events were really understood. It is the causes behind the causes that have eluded political theorists. Those causes were previously inaccessible to the state of knowledge and are now only beginning to be reachable through the biobehavioral and other sciences. Humans are affected by forces external to themselves, but those effects are manifest in the reactive capabilities of the human mind/body organism.

Not everyone would concede the practical significance of such deep-seated innate causal relationships. Their existence could (not necessarily would) threaten a very broad range of social, political, and moral theory. Denials that they exist or of their practical significance are sometimes uttered with a vehemence that suggests a fear that some degree of biological determinism *may* exist and *could* contradict strongly held emotional commitments. Claims of genetic determinism unsupported by demonstrable evidence reinforce skepticism. In fact, the present state of biology offers very limited insight into such concepts as social health and sustainability which must surely require a synthesis of internal and external factors.

Substantially more biobehavioral information exists than is applicable in public policymaking, but prospects are promising for a much more adequate understanding of human behavior through advances in the sciences of life and environment. Nevertheless, the prospects of this knowledge being put into forms suitable for public choice and policy-making are at present much less likely. Given the social discontents and disorders of our times, there is reason to suspect that society today is not proceeding on premises which are valid. There are more wrong than right answers to any problem and right answers do not follow from false premises. If its assumptions are invalid, our present society may not indefinitely survive miscalculations of its own making. We do not know how much disorder or disaffection our present forms of society will stand.

One need not invoke doomsday rhetoric to make the point that human societies have often been and have continued to be misguided by their own misunderstandings. Humans have only recently begun to study their behavior with methods and instruments that can penetrate beneath surface manifestations. We have emphasized Claude Bernard's concept of human behavior resulting from interaction between man's *milieu extérieur* and his *milieu intérieur*, regarding it as indicative of the need to achieve some practical linkage, if not integration, between the biological, environmental, and social sciences.

To understand the behavior of organisms both aspects of environment and their interactions must be understood. It has been less difficult for humans to study and objectify the external aspects of their environments than to discover the relevant aspects of environmental relationships within themselves. In a wholly rational world, investigation of these environmental components and their interactions would be of highest priority. In the present world, the configuration of knowledge does not lead to this allocation of research priority; science is not well organized to approach reality from this viewpoint which is multidisciplinary and interdisciplinary. Our commonsense view of environment still places it outside the environed human being—as an externality.

Changes in behavior attributable to advances in biological knowledge and application are directly and indirectly so diverse as to make reliable projections of possible consequences very difficult. But no great effort has been made to develop such projections. With the present state of knowledge and technique, forecasting social futures should provide estimates of where at least some biological trends might lead. Legislation introduced into the 99th and 100th Congresses of the United States would provide means toward developing this capability.[11] More than biocratic trends would be studied, but the power and potential of the new biology ought to make its interactions with other aspects of social change a major object of attention.

A Brave New World?

The more celebrated conjectures regarding mankind's biological future have been generally regarded as unattractive. For readers of Aldous Huxley's *Brave New World* (1932) this understates the reaction of many who see in advancing bioscience a foreshadowing of the dehumanizing Huxleyan "utopia." News media reports of "test-tube babies," genetic screening, prenatal sex selection, cloning, and recombinant DNA technology add up to prevision of a biocratic world in which individuals are regarded primarily as receptacles for the transmission of genes. Popular interpretations of sociobiology such as Richard Dawkins' *The*

Selfish Gene (1976) could reinforce those fears. Regardless of its potential for the alleviation of human misfortune, there are efforts being made to closely control or even to ban research in human genetics.

Fear that research on human genetics will lead to eugenics which will lead to social Darwinism and the racial practices of National Socialist Germany reflects a distrust of applied science that may be misguided but is not wholly irrational. Misuses of science, intended and inadvertent, have occurred too often to justify a complacent attitude toward biological innovation. There appears to be greater popular sensitivity to the hazards of applied biology than to risks from other areas of applied science except perhaps with regard to exposure to nuclear radiation and to dangerous chemicals, both of which threaten human life and may cause transmittable genetic damage. A major contribution of biology to human welfare is the identification of those elements in the external environment that impact upon human life and health.

Biology is concerned with more than genetics and the internal environment of organisms. The fitness of the environment for life has been a continuing object of multidisciplinary scientific inquiry. Life is the focus of concern, but more than biology must be involved to understand the conditions necessary to life on earth and possibly elsewhere in the cosmos. Living material, not human society, has been the focus of most of this inquiry, yet today human society impacts upon all living material.

The social sciences have rarely been involved in collaboration with the life and physical sciences in comprehensive and continuing efforts to understand human behavior in an integrated social-ecological context. Environmental influences have been extensively invoked to explain human history, but the widely prevalent view of man as an autonomous independent creation, outside of nature, has discouraged serious inquiry into the biology of human social behavior. The science of psychology has an obvious concern for physiological influences but its focus tends to be on individual and interpersonal behaviors that are accessible to available methods of scientific investigation. Under the present configuration of knowledge no single science is adequate to the task of understanding human social behavior in its biospheric context.

We are not yet near an adequate science of human behavior and have advanced little beyond speculation regarding the possibility of such a science and how it might be developed (assuming, as not everyone does, that this development would be desirable). The sciences of anthropology, psychobiology, and sociobiology afford respective avenues toward such a development but have not taken us very far toward the understanding of what humanity needs if it is ever to control that part of its destiny that is humanly controllable.[12] There are many aspects of human behavior and society that are presently beyond the boundaries of science. Some may never be accessible to science as we know it today. Yet regardless of fears and objections, advancement of the life sciences seems as likely as any forecast that we can make. The argument for more attention to its implications is persuasive to those persons who conceptually bridge the life sciences, the social sciences, and the humanities. Their numbers, however, are relatively few and they are not in step with the present academic preference for analytic technique, or with widespread disregard for substantive and institutional investigations that are not amenable to quantitative evaluation. Discipline-focused methodological rigor is not the exclusive goal of academic researchers today, but is dominant. Meanwhile, the interdisciplinary field of the life sciences and public policy receives much less attention than its intrinsic importance merits.

A strong case may be made for more attention to public policy and the life sciences, but the argument does not wholly assert the need for more research.[13] A case may also be made on the need for more attention to what is already known and for a logical and informed examination of the basis of prevailing attitudes, beliefs, and behaviors. If policies are being adopted upon the basis of assumptions inconsistent with the parameters of the real world, persistence in error will ultimately lead to political checkmate, possibly to social failure. The history-indifferent, highly individualist Western society of today is poorly prepared to examine the long-range social consequences of its premises, tacit or explicit. If a subliminal unease about the future characterizes the more thoughtful members of this society, the feeling may shortly prove to have been justified.

The basic assumptions upon which the social and economic policies of all modern states are based appear to be contradicted by the limits of the natural world which encompass all that is human. The modern world proceeds on assumptions that are mechanistic rather than organic, linear rather than multidimensional, even though science since Darwin and Einstein presents a more complex interpretation of reality. Newtonian logic, not ecology, continues to provide the conceptual mode of political thinking. Future historians may regard as astonishing that more than a century after Bernard, Darwin, and Mendel, no coherent political philosophy had displaced Marxist interpretation of political behavior.

The key assumption of modern political-economy is the necessity for continuous growth. There are various interpretations of what growth means, but political rhetoric is characteristically unspecific and uninformative. Growth, in the language of economists, does not always carry the same meaning—and too often the meaning is misunderstood by the ordinary citizen. To the economically unsophisticated, which includes almost everyone, growth means what it seems to mean, the indefinite expansion of population, capital, technology, material goods, and opportunity for advancement in the social hierarchy—capitalist or socialist.

The gospel of growth has encountered critics who have pointed out the unwanted consequences of an indiscriminate commitment to the proposition that bigger must be better. Evidence that indiscriminate growth may destroy the basis of its continuation have prompted governments and their economic advisors to seek policies for "sustainable" growth. But what kind of growth is sustainable in a finite world? To believe that exponential growth of any kind is sustainable is to disbelieve reality as revealed by science. Indefinitely sustainable growth appears to fall into the same category of concepts as perpetual motion and no more rational as a premise for public policy.

The apostles of growth may not believe that all forms of growth are indefinitely sustainable, but which are and which are not are seldom specified. Political economists usually tacitly assume a foreshortened time horizon for their projections—not often more than a quarter century into the future. Limits to growth studies have been severely criticized for going beyond hard evidence toward an unpredictable future. If growth in an expansive sense is not indefinitely sustainable, and if the welfare and sustainability of society nevertheless depends upon its continuity, the question "what then" arises if growth in an expansive material sense ultimately must end.

Democracy must also contend with more deep-seated conceptual difficulties that the life sciences might help to clarify. This is not the place to critique the moral basis of liberal democracy but, having suggested that some of its assumptions may be questionable, there is an implied obligation to point out what the more serious of them may be. It may be that the assumptions could be interpreted in ways consistent with biocratic realities, assuming that these realities could be apprehended. Lacking that grasp of basic knowledge, society adopts contradictory or mutually counteractive policies through failure to reconcile basic beliefs. No one has done more than Garrett Hardin to expose the fallacies of social dogma that are contradicted by biological realities.[14]

For example, common beliefs regarding human equality and diversity necessitate political fictions to justify an implicit assumption that, in principle, each individual should have an equal voice in political decisions, ergo, one person-one vote and that there is an obligation to exercise that right. Democratic equalitarian principles are reflected in demands for racial and social integration but conversely human rights are asserted on behalf of ethnic identities which to survive require some degrees of segregation. Tacit assumptions regarding human intelligence characteristically fail to take account of intellectual diversity, although differences are commonly acknowledged in practice. In many societies certain rights are regarded as inalienable yet are systematically disregarded under particular circumstances. Yet rights are more readily conceded than responsibilities enforced. Liberal democracies have shown much less success in internalizing a sense of responsibility in their members than in sensitizing them to their assumed rights and liberties.

Biology becomes relevant when rights are asserted in relation to numbers of people. The moral philosophy of Western liberal democracy provides little protection for its survival against submersion beneath whatever cultural or political changes accompany the outflow of people from rapidly expanding populations of the so-called Third World. It is conceivable that Western liberal democracies could be overwhelmed in the name of their own principles, universally construed. Preventive measures, however, are sufficiently distasteful to many persons of humane and liberal persuasion that the response to the perceived threat is often a rhetorical "would that be too bad?"[15] It is not clear that under the assumptions of contemporary liberal democracy that any people or nation have a "right" to defend their ethnicity or standard of living against the "needs" of their fellow humans even though those needs may have been the consequences of what might be described as biological folly.

In brief, the thesis of this book has been that the critical and seemingly intractable problems of modern society may be significantly affected by misunderstanding of their causes and possible consequences. This failure of understanding is suggested by social and political commitments to propositions and doctrines that are not demonstrably true and hence may be false. The argument is not that the assumptions underlying liberal democracy as presently understood *are* wholly invalid, it is rather that the conditions of society and governance strongly suggest the prevalence of dangerous misconceptions of the basis of human behavior and survival. In brief, there is reason to suspect the presence of serious faults in the foundations of modern society.

If there is reason to suspect that misconceptions exist, it would seem prudent to identify what they may be and to ascertain their causes and cures, an argument advanced by biologist Bernard D. Davis in an essay on "Novel Pressures on the Advance of Science."[16] This effort would surely lead to discovery of the gap between the sciences of life and the sciences of society. It would also reveal a domain of the humanities that is weakly connected with any of the sciences except in respect to certain aspects of technology.

It is emphatically *not* the argument of this book that the life sciences per se will or can solve the problems of society or that traditional values, unverified by empirical tests, should be assumed to be inconsistent with nature. Science is a human artifact and provides no infallible guide to conduct or policy. It may, however, inform human choices and expose assumptions that lead to folly. The argument thus implies that the social and policy sciences including the law should give greater attention to what may be learned from biology and find ways to strengthen linkages between the biological and social sciences and the humanities. Ultimately, the linkages should lead to integration so that the unity of science becomes a reality as well as an ideal.[17]

The risk of learning from the life sciences is less than the danger of failure to learn. The prospect of a *Brave New World* is greater through failure to anticipate and rationally to address the effects of advancements in the life sciences than to evade consideration of their implications. Biocracy and democracy can be allies if democratic societies give informed and timely attention to the findings of biology and its applications in biotechnology. No less important is support for an integrative interrelating of the biological and social sciences in the structure and procedures of higher education.

A dogmatic, inflexible interpretation of democracy will in all probability lead to ultimate loss of values sought through democratic governance. More than any other form of government, democracy depends for its vitality and survival upon informed and responsible conduct by governors and governed alike. Through science, society has acquired unprecedented power to influence its future. Without serious regard to the relationships between the conditions and processes of life and of human behavior, it is hard to see how society can safely assume responsibility for that future. To discover and deal constructively with these relationships, a convergence of biological and social sciences and the humanities is necessary. Nothing less than a fundamental reexamination of the assumptions upon which modern liberal democracy depends may be required to provide it with a basis for survival in the years ahead.

Notes

1. Arnold J. Toynbee, *A Study of History*, 12 vols. (New York: Oxford University Press, 1948–1961).

2. At one of the last opportunities for the Aztecs to drive the Spaniards from Mexico their emperor Montezuma, under duress, addressed his subjects, according to Prescott's account, saying, "Have you come to drive them from the city? That is unnecessary. They will depart of their own accord if you open a way for them. Return to your homes then, lay down your arms. Show your obedience to me who has a right to it. The white men shall go back to their own land; and all shall be well within the walls of Tenochtitlan." William H. Prescott, *History of the Conquest of Mexico*, vol. 2 (Philadelphia: J. B. Lippincott, 1853), 307.

3. See the journal *Politics and the Life Sciences* and for numerous relevant publications not directly cited in this book, see two interesting and informative treatments of scientific ideas and politics: W. J. M. Mackenzie, *Biological Ideas in Politics: An Essay on Political Adaptivity* (New York: St. Martin's Press, 1979); and E. M. Friedwald, *Man's Last Choice: A Survey of Political Creeds and Scientific Realities* (New York: Viking Press, 1948).

4. Ralph W. Gerard, "Higher Levels of Integration" in *Levels of Integration in Biological and Social Systems,* ed. R. Redfield. (Lancaster, Pennsylvania: Cattell, 1942), 67–87. But note also Alex B. Novikoff, "The Concept of Integrative Levels and Biology," *Science* 101 (1945): 209–15.

5. Jay Forrester, "The Counter-Intuitive Behavior of Social Systems," *Technology Review* 73 (January 1971): 53–68.

6. Paul W. Taylor, *Respect for Nature: A Theory of Environmental Ethics* (Princeton, New Jersey: Princeton University Press, 1986).

7. For a comprehensive interpretation of societal and intellectual evaluation consistent with the biocratic orientation, see Peter A. Corning, *The Synergism Hypothesis: A Theory of Progressive Evolution* (New York: McGraw-Hill, 1983).

8. David Joravsky, *The Lysenko Affair* (Cambridge: Harvard University Press, 1970); and H. J. Muller, "The Crushing of Genetics in the U.S.S.R.," *Bulletin of the Atomic Scientists* 4 (December 1968): 269–71.

9. Edmund W. Sinnott, "The Biological Basis of Democracy," *Yale Review* 35 (December 1945): 61–73. On the behavior of the individual as a biological entity, from a philosopher's perspective, see Hans Jonas, "Biological Foundations of Individuality," in *Philosophical Essays: From Ancient Creed to Technological Man* (Englewood Cliffs, New Jersey: Prentice-Hall, 1974). That the biological basis of individuality may have an explicitly political significance is suggested by work published by Douglas Madsen. See "Power Seekers Are Different: Further Biochemical Evidence," *American Political Science Review* 80 (March 1986): 261–69, and "A Biochemical Property Relating to Power-Seeking in Humans," *American Political Science Review* 79 (July 1985): 448–57.

10. Cf. G. Leach, *The Biocrats* (New York: McGraw-Hill, 1970). Leach's book is primarily a study of biomedical ethics, not of biocracy as the term is used in these essays. On the meanings of democracy Harold J. Laski wrote, "No definition of democracy can adequately comprise the vast history which the concept connotes" (*Encyclopedia of the Social Sciences,* vol. 5 [New York: Macmillan, 1931], 76).

11. S. 2905. Mark Hatfield, "Global Resources, Environment, and Population Act of 1986." A similar bill was scheduled for introduction into the House of Representatives on 29 April 1987.

12. E.g., Ernest Becker, *The Lost Science of Man* (New York: George Braziller, 1971); Karl William Kapp, *Toward a Science of Man in Society: A Positive Approach to the Integration of Social Knowledge* (The Hague, Netherlands: Martinis Nijhoff, 1961); and, Laura Thompson, *Toward a Science of Mankind* (New York: McGraw-Hill, 1961). See also *Psychobiology: The Biological Bases of Behavior,* Readings from the *Scientific American* (San Francisco, California: W. H. Freeman, 1966); and, of course, more recent publications in this field. Sociobiology is extensively referenced in chapter 7.

13. Note earlier referenced report (chapter 1, note 23) on *Biotechnology, Public Policy, and the Social Sciences.*

14. His writings are numerous and are perhaps best referenced to the card catalog in a good library. However, see *The Limits of Altruism: An Ecologist's View of Survival* (Bloomington: Indiana University Press, 1977), and *Naked Emperors: Essays of a Taboo Stalker* (Los Altos, California: W. Kaufmann, 1982).

15. It seems fair to say that most members of the Western intelligentsia view with disdain "survivalist" writings such as Jean Raspail, *The Camp of the Saints* (New York: Scribner, 1975); Robert and Leona Rienow, *The Great Unwanteds Want Us* (1980); or Wilmot Robertson, *The Dispossessed Majority* (1976). These are not "scientific" books; but fiction and advocacy writings may make plausible statements where "respectable" social science is unwilling to tread.

16. Bernard D. Davis, "Novel Pressures on the Advance of Science," *Annals of the New York Academy of Sciences* 265 (1976): 193–202.

17. Note e.g., Carl G. Hemple, "General Systems Theory and the Unity of Science," *Human Biology* 23 (December 1951): 313–22. For additional references see Lynton K. Caldwell, "Ars-Techne-Scientia: Integrity of Culture Needs Unity of Knowledge," *The New Liberal Arts—Curriculum in Transition* (Albany, New York: State University of New York Press, 1985): 7–14.

For Product Safety Concerns and Information please contact our EU representative GPSR@taylorandfrancis.com
Taylor & Francis Verlag GmbH, Kaufingerstraße 24, 80331 München, Germany

www.ingramcontent.com/pod-product-compliance
Lightning Source LLC
Chambersburg PA
CBHW060742150926
51501CB00003B/69